Ken Roberts is a workplace cultural journeyman with over four decades of experience across mining, heavy industrial and construction. His career encompasses roles as a construction and safety manager, procedure and systems project manager, and he is the founder of Transfoorm™, a cultural transformation consultancy for the biggest mining companies in the world. His extensive real-world experience enriches his provocative writing, providing practical insights and valuable tools. Ken's passion for personal development, coaching and facilitation, puts a unique perspective on workplace safety and culture, making his work compelling for those seeking a fresh approach to engagement, productivity, morale and retention. He blends workplace training with personal development techniques, offering readers practical solutions laced with thought-provoking ideas and a distinctly novel approach.

Theo Venter is a motivational speaker renowned for his impactful presentations on workplace safety. Surviving a life-threatening accident involving 22,000 volts through the heart, Theo shares his experience to transform safety culture in workplaces around the world. His approach focuses on emotional connection and taking personal responsibility for safety. His story and insights are a powerful catalyst for change, making him a sought-after speaker and expert in global workplace safety and cultural tran

CONVENIENCE KILLS

The Mystery Cause of All Workplace Accidents Revealed!

Ken Roberts | Theo Venter

Transfoorm

CONVENIENCE KILLS

The Mystery Cause of All Workplace Accidents Revealed!

By Ken Roberts and Theo Venter

Enquiries:

- ken@transfoorm.com (+61 0415 356 357)
- www.transfoorm.com
- www.conveniencekills.com
- www.au.linkedin.com/in/thekenroberts
- www.au.linkedin.com/in/theo-venter
- www.theoventer.com

Book cover and design by Transfoorm™

(241008)

ISBN: 978-0-6485086-2-5

10 9 8 7 6 5 4 3 2 1

Convenience works everywhere
in the world... except safety!

Special thanks to Kenneth Manyumwa from BHP Olympic Dam. He inspired many innovations through his willingness to try new things with his work teams and was the the first to prove the positive effects that "Convenience Kills" has on safety and productivity.

Contents

Introduction

Theo Venter

Bloody Mondays!!?

At first light we are told there's a change to our routine, an urgent repair to a high-voltage transmission pole damaged by lightning overnight. Our three-man crew arrive and look up to see twisted metal and wire. A direct hit, what a mess!

We start breaking the news to businesses that we need to isolate the street's power—except one of them complains, *"What, turn off my power? I've got a business to run."* So, to avoid confrontation we prepare for a "hot-line." Sure, there are 22,000 volts of electricity buzzing around, but no problem, we do this all the time. It's *Just Another Day*!

Hold tight; before I get ahead of myself, don't assume this will turn into one of those spine-tingling sagas designed to scare you into always following safety. It's more of a mystery tale than a horror story. What I do next has left even the most seasoned safety professionals dumbfounded, not to mention myself. But I hesitate to reveal the plot on the first page, so let's kick off with why Ken and I wrote this book. It's been quite a journey.

Perhaps you're familiar with our first book, "GET REAL - Staying Alive for a Living." In 'Get Real' we meticulously deconstructed everything leading up to and after my accident. However, this book narrows in on that one pivotal moment when I made a Convenient choice which would change my life forever. Because frankly, people demanded it.

Spoiler alert: Convenience "IS" the underlying cause behind most, if not all, workplace incidents and accidents. There you have it. It's done. The cat is out of the bag.

But is it? Has the full picture truly been revealed? How has this thing called 'Convenience' remained so elusive to health and safety professionals for so long? It's more than just 'complacency' or taking a 'short-cut.' Convenience is so seductive that it flies under your 'safety radar' every day.

Our research reveals a gaping hole: hardly anyone is addressing the impact of 'Convenient choices' as the root cause of workplace incidents. The chapter called Convenience Kills from 'Get Real' resonated deeply, affecting many in ways no other chapter did. People were saying it improved safety. So obviously, it's now my mission to spread the news.

If you've seen me present at your workplace or at a safety conference somewhere, or you visited www.theoventer.com, you know my story. It's harrowing, to say the least. Surviving a jolt of 22,000 volts directly to my heart is nothing short of a miracle. It's fate. Kind of how I met Ken.

After one of my talks, he approached me, insisting we have a chat. I was reluctant because I had to leave immediately for my next gig on a remote mine site in the middle of nowhere. To my amazement he was there too on a completely different assignment. Coincidence? I don't think so.

Our first conversation was profound. He basically came out and said, *"I know exactly why you had your accident, I've discovered the "real" reason—it's Convenience."* Bang! His insight was spot on—connecting the glaringly obvious "Lure of Convenience" to my accident. While Convenience wasn't new to me, its deep-rooted link to my accident was. He went on to explain how Convenience was deeply ingrained human behaviour and that many others are bound to suffer the same fate if we don't do something about it. So we shook on it.

Sadly, these sort of safety discoveries are often turned into sermons, preached to us like in church. So we vowed to change that. We're offering a completely different perspective, breaking it down, so you experience my story and realise its relevance at a deeply personal level.

Ken's writing captures my ordeal perfectly. Through his words, you'll comprehend the gravity of the decision I made that nearly killed me. It should have. Therefore, when Ken writes "I", it symbolises all of us. We're in this together.

It's great to be able to share my story through Ken's writing. I urge you to dive into his narrative and see safety through his eye's. It's new and innovative. There's no foreword or preface. No acknowledgements. Just revelations. We offer you unvarnished, invaluable facts. The root cause of workplace incidents? Well, yes, it IS Convenience. And I'm living proof. So let's unravel this mystery shall we?

But whatever you do... please take care out there!

Panel of Experts

It's getting hot up here. Just get on with it will you?

Theo's crew run through their safety checks and complete the risk assessment, throwing the paperwork into the front of the truck. The two more experienced *lineys,* Theo and Nico, don their Personal Protective Equipment (PPE) and jump into the cherry picker. They raise it up and begin repairing the damaged insulator leaving the other crew member on the ground as the safety observer. 22,000 volts of electricity buzz around them like it has dozens of times before.

By mid morning things aren't going well. Up in the basket, Nico struggles to remove a stubborn 12-millimetre nut. How are you supposed to get your fingers inside a thin metal channel wearing thick heavily insulated protective gloves? Who designed this stupid thing? Not only is the sun heating up, but so are frustrations.

Eventually, Nico gives up and asks Theo to have a go. Yet after a while, Theo is stuck too. There's just no way to do it. Except, failing is not an option for Theo. It's not in his make-up to quit. Everyone relies on Theo to get the job done. If a job gets tough, Theo gets going. He's Mr Fix it, the Go-to guy. Why should this time be any different?

It's about now that the "voices" in Theo's head take over. Like a panel of experts out of nowhere, they start to toss a few ideas around. One voice says, *"Let's lower the basket and call the boss."* But immediately there's a retort, *"What?, give up, no way."* Theo's *experts* decide that this would be akin to admitting failure. So another voice suggests, *"Maybe we should go back to the yard and find a special tool."* But that is met impatiently with, *"Forget it, too long."*

Let's stop the story for a moment. Can you guess what is going on? Exactly. Theo's panel of experts are weighing up if any of these suggestions are going to Inconvenience things further. As if further slippage in the schedule is going to have some sort of irreversible impact.

As it turns out, this is a very human thing to do, occurring unconsciously, taking possession of our senses. Evaluating whether our ego is exposed. Much more than if our physical being is. Remember, Mr Fix-it is in control so a bruised ego is definitely Inconvenient.

I guarantee you've been in this situation too. Who hasn't? It's Monday. You drop everything. Go to another job. Bang heads with a stubborn customer. A stupid 12-millimetre nut is derailing the schedule. What else can go wrong?

Well, a million things really. This is the perfect storm. A template for tragedy. I've never investigated an incident, and there have been hundreds, where there wasn't a host of distractions occurring simultaneously. I call it "in the heat of the moment."

Accident causation academics will tell you it's the Swiss cheese model. Well, maybe. But throw in some ego, frustration, dented pride, conflicts of interest, and you have a recipe for disaster. Let alone time constraints, budget concerns or production targets? The list is endless. When we add irrational human-beings into the mix, causation models probably oversimplify things. You only need one or two simultaneous distractions and it sets off a chain reaction that has you searching for an immediate solution. One that seems to be needed instantly. Whether a solution is really needed or not right now—is not even considered.

Our brains have a remarkable ability to rationalise the choices we make, regardless of ethical or logical considerations. This phenomenon speaks to our deeply ingrained desire to justify any actions as 'right' in order to maintain a sense of self-consistency and avoid mental conflict. It's crucial to understand the underlying cognitive biases that influence our decision-making in high-pressure situations.

Meanwhile, back to the story.

Nico is turned around, taking in the view and talking with the spotter below, waiting for Mr Fix-it to do his thing. He's oblivious to the turmoil going on inside Theo. And all this is happening in the blink of an eye. Barely nanoseconds pass.

At All Costs

Whatever decision you eventually make,
needs to seem like it was the right one.

Theo's *panel of experts* continue to debate what to do while he ponders how on earth other work crews have been doing this. He recalls seeing a guy remove his insulated gloves once and working bare-handed. Theo doesn't know if it was for this particular nut problem or something completely different. At any rate, the idea is out. Immediately one of the *experts* latch onto it. *"Hey that's a great idea."*

Although Theo had never taken his gloves off before, this would surely work. But it's going to take a little more justification than being Mr. Fix-it or the Go-to guy. If the boss found out that Theo removed his PPE, he'd be dismissed on the spot. So that's out of the question.

Yet this is where it gets interesting. Our brains always reach for the most Convenient option in these moments and will rationalise its way through anything. I will explain more in detail soon, but your brain is hard-wired to find the easiest solution when things *heat up* and will find dozens of ways to make sure it was the right choice.

Exasperation, dissatisfaction, weariness, disappointment, aggravation, heightened senses generally—you name it—will trick you into believing that these states require your immediate attention. Convenient choices become very attractive right now to conserve brain power and save energy. Because energy and brain power is being burned furiously when you are in these conditions. And our brain hates that.

To seal the deal, your team of voices will rationalise whatever decision you arrive at. We've been raised that we must make good choices and avoid bad ones. So whatever decision you eventually make, needs to seem like it was the right one. Whether it 'IS' right or not doesn't seem to matter. Just how far can a human brain go to normalise an abnormal situation. Let's see shall we?

"What the hell," Theo considers. *"I've been in this game for 17 years. I've forgotten more than most lineys know. I know what I'm doing. I'm the best liney in this goddam company."* Then fatefully the strangest thought enters his mind, *"Right, I've got to loosen this nut at all costs."*

Let's hold things right there. *"At all costs."* Really? Has Theo thought this through? Consider where his mind is right now. He is totally 'IN' the moment. Hooked! The minutia of the nut has his laser-like focus and attention. As far as Theo is concerned there's only one thing happening right now in the world.

You've been there too, yes? Think when you were in this

trance-like state. Time comes to a standstill and space vanishes. Everything around you is blocked out. There's only the nut. This is human nature at its finest. Our biology is designed to conserve power and store it for sudden physical bursts of energy required to catch prey or escape predators. Our *Primitive* brain doing its job.

Back in the basket, Theo makes a decision to reduce the intensity. He clamps his gloves between his knees and withdraws his hands—quietly, so Nico doesn't catch on. He reaches into the channel and has free access to that stupid nut. It makes the job so much easier. The nut practically falls off. Theo feels a sense of relief, if not a tinge of pride. He's beaten the bloody thing!

Then suddenly, without warning, the whole cross-arm begins to fall. Theo senses immediately that his workmate below is about to get clobbered by a heavy metal falling object. A chemical concoction of adrenaline is sent instantly racing through Theo's body as a reflex action. Unconsciously... instinctively... he lunges out to catch it!

It's about now that Convenience, Swiss cheese models, root cause analysis or any other explanations go right out the window. Humans are human. Instantaneous snap decisions, like reaching out to catch something from falling is an automatic bodily response. A reaction stored away in a bionetwork of neurological pathways. Deeply rooted in fight or flight responses of the brain. No scientific explanation or words from a book will influence a human in this mode.

Alas, poor Theo is on his own.

Just Get The Job Done

Whether it's right or wrong doesn't seem to matter.

Theo's left wrist makes contact with the power line. 22,000 volts rip through his body, as if a whispered curse from hell was sent to end his being with one searing bolt of lightning. An intensely fierce celestial fire consumes his entire body. Raw energy roars through his veins. Every muscle contorts in a massive, distorted spasm.

> *"I felt everything. I remember each and every second,"* Theo recalls. *"People say you block it out, but you don't."*

The electrical force, looking for an exit point, rushes to Theo's right wrist which was resting against the metal cross-arm. His left hand which contacted the power line is instantly disfigured for life. At two and a half seconds, Theo breaks contact with the beast and falls back into the basket at Nico's feet. Nico thought Theo was dead, but kept his cool. All those years of training had primed him for this. He lowers the basket as fast as he can, with Theo's body smouldering and contorted on the floor. Unexpectedly, Theo regains consciousness. His blood is literally boiling. Bones and flesh are smoking. Theo panics at the stark realisation that he is still alive. Then, *the screaming starts.* On the ground, cold water flushes and first-aid do nothing to ease the agony. Theo is literally cooking from the inside. In the hospital, doctors slice open Theo's arms to relieve the massive inflation. Antibiotics are pumped into him by the truckload.

So, let's take a breath. Obviously, not all accidents are going to end up like Theo's. But the processes leading up to Theo's accident are no different than any other. If we undertake a deeper exploration into all the reasons Convenience, or the avoidance of Inconvenience, are truly the root cause of accidents, you will see a pattern emerge in your own workplace.

For years I've interviewed work crews right after incidents, while their memories are fresh. After gaining their trust, we discuss their thoughts just before the incident. Consistently, a dominant thought emerges: "Come on, just get this job done." This urgency overrides any safety training or procedures. Decisions snap to the most convenient option or shortcut, disregarding whether it's safe or not. Thoughts of potential impact, long-term consequences, or injuries are nowhere to be found.

As Theo slips his hands out of his gloves, there is no flashback to the risk assessment, no recall of the promises he made to his family that he would take care of himself... nothing. It's a phenomenon really, and confounds safety professionals no end. The awareness that things might not turn out so well is overridden by an incredible impulse to finish the task. When you are confronted with a work task challenge, and there is no procedure or risk assessment guiding the next move (for whatever reason), it will depend on what the workplace culture is, as to how things will pan out next.

If there's a high value around job completion and production, you are likely to experience tension during challenging work tasks and want to *get the job done* at all costs. But if you are guided by a sense of "teamwork and camaraderie," then you will feel more confident to do whatever it takes to be safe, for the team's sake perhaps more than your own. Especially if the workplace has defined teamwork as providing more fulfilment than personal achievement and safe-work decision making more desirable than performance.

This is so critical to understand because it's the single biggest mistake we are making in our workplaces. With no apparent *purpose* that encompasses a vision for what the workforce believes is "their" desirable environment to work in (i.e. Fair or Respectful etc.), there is no underlying "Catch-All" behaviour that underpins *heat of the moment* decisions. There's no "back-up" action that might otherwise be triggered for situations that fall between the cracks of unfollowed procedures or processes. For no matter what, if our workplaces include humans, there will be cracks.

Theo's electrifying brush with death, inflicted by a staggering 22,000 volts, offers more than just a tale of personal survival and agony. It presents a crucial lesson in the essence of effective workplace safety that extends beyond the mere adherence to protocols. His experience sheds light on the pervasive yet subtle peril of prioritising Convenience over caution, a common yet overlooked cause of many workplace accidents.

In my interviews with teams after an accident, a recurring theme surfaces: in the crucial moment before an accident, in the 'heat of the moment', the overwhelming pressure to "just get this job done" often silences the disciplined echoes of our procedures, the voices of our training, and especially behavioural based safety and risk assessments. This urgency leads to choices favouring speed and ease over safety, sidelining potential risks and long-term consequences. Theo's story hits hard, and these images hit even harder. They are more than pictures; they are a wake-up call. They silently shout that safety isn't a checkbox or a protocol, it's a 'gut feel'.

Convenience is like a wolf in sheep's clothing. We 'tick and flick' our way through shortcuts, bypassing safeguards, only to scrape through with a close call or scare. We rarely face serious consequences. Yet, this is the moment of truth: is the shortcut worth the risk?

Every choice has weight, every shortcut a consequence. How do we make workplaces understand that 'safety' isn't just a sign on the wall, but a promise we keep to ourselves. Let Theo's accident help you to consider what Convenience is capable of. Let's get real—because the cost of not doing so is etched in these images on the next page.

We might consider going out into the field next time with safety in our hearts, and leave productivity KPI's and Lean Board requirements in the Pre-Start room where they belong. Lest they urge us to 'get the job done'. Convenience Kills!

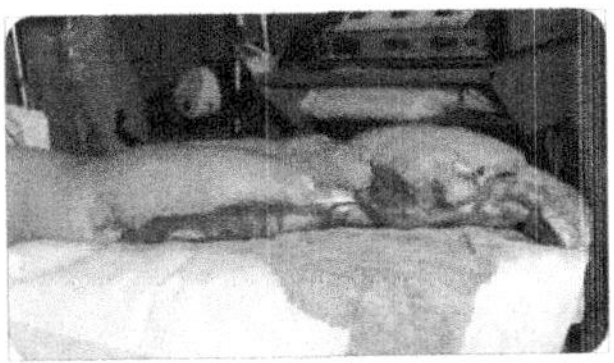
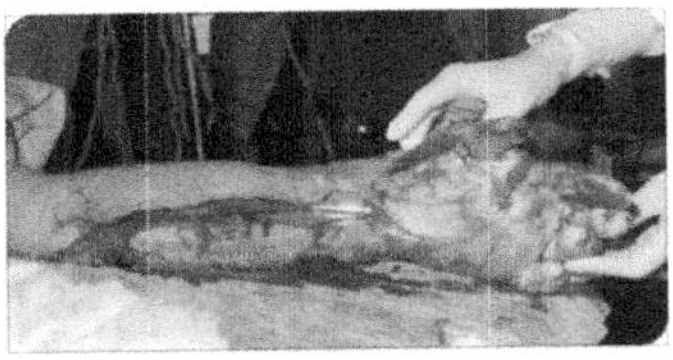
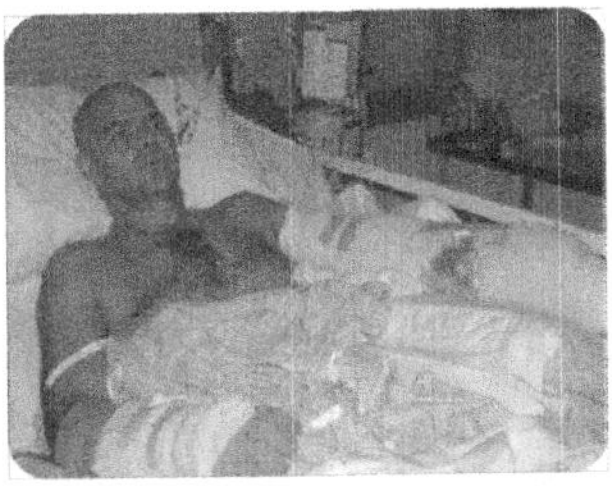
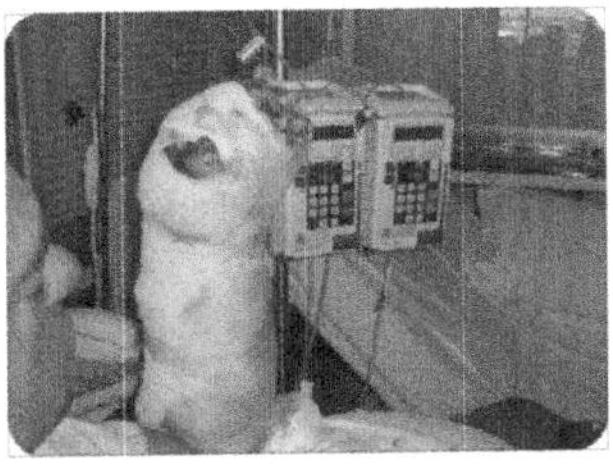
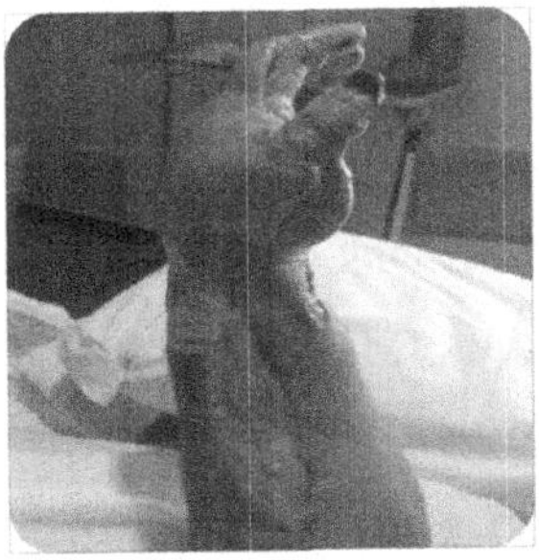
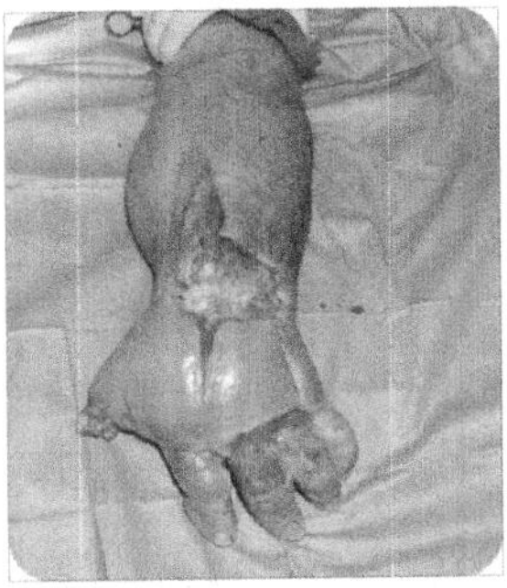
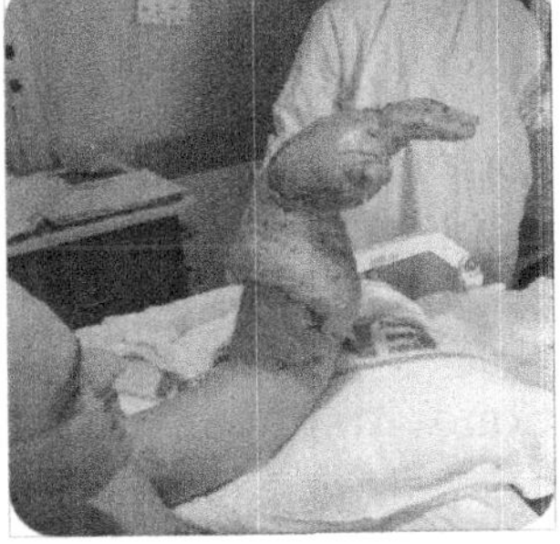
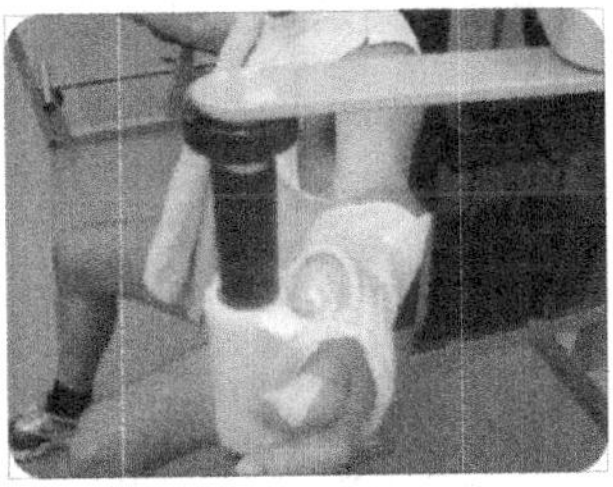

The Lone-Wolf

Theo could have elbowed Nico in the ribs and said, "Hey should I remove my gloves?"

The original safety controls in the risk assessment were sitting in the front of the truck. The furthest thing from Theo's mind. Instead of slowing things down, he sped things up. Compressing the situation and creating urgency. But like a release valve on a pressure cooker, something had to give.

Theo's 'panel of experts' started rationalising, "Maybe we tell the customer there's no choice but to isolate the street's power." "Nah," says a counter-argument, "Who wants confrontation?" Then from out of nowhere, "Bloody hell boys, it's getting hot up here, just get on with it will you!"

This tone caught Theo's attention. A distant parental command perhaps: "Hurry up, pick up your toys, stop messing about!" Wherever it came from, it evokes corresponding body feelings; stress, frustration... fear! This tension makes Theo believe he has to do something to relieve the situation. But actually, there's no need for him to do anything. Nico is right there, ready to help. Theo only needs to ask.

Caught in his own psychological dilemma, triggered by buried unconscious experiences of his past, this familiar pattern compels Theo, almost against his will, to take some kind of action. It's about now we want that "Stop Work" campaign to kick in... and also why it didn't.

We rarely consider these root causes in our incident investigations: stress, tension, ego, psychological disposition. These states urge us to make decisions we wouldn't under calmer circumstances. Conquering difficult problems serves as a coping mechanism for people like Theo. Getting the job done is a familiar pathway in dealing with complex issues. At least that's how he's dealt with them in the past. Even rewarded by his peers and management.

The biggest takeaway here is the profound impact emotional states have on our actions, particularly in high-stress situations. Stress, tension, and agitation are experienced not just as feelings but as active agents that compel us toward hasty actions—often against our better judgment.

What happened to Theo serves as a cautionary tale about the perils of caving into emotional compulsion and the value of seeking counsel rather than succumbing to the insidious trap of self-reliant problem-solving.

It invites reflection on the importance of collaboration and the dangers of lone-wolf tendencies in high-stakes environments. This tale warns of the human tendency to go it alone and the importance of developing the power to seek support, especially when one's instinctive responses are influenced by past experiences and subconscious triggers.

It's crucial to understand that Theo had another option—through his colleague Nico. Why didn't Theo just elbow Nico in the ribs: *"Hey Nico, should I remove my gloves?"* We don't know what would have happened. But Nico represents the missed opportunity of a lifetime—in not asking him for advice. We'll never know what could have been, but in a high risk workplace, problems should always be solved together, never alone.

Theo, trapped in his habit of trying to fix everything himself, ignores this choice. He sticks to the belief that he must solve the problem on his own, without any help. This frozen moment in time underscores the human propensity for self-sabotage: the inclination to exacerbate rather than alleviate our anxiety, particularly when influenced by the muddy waters of our subconscious.

It highlights the imperativeness of not merely being aware of one's emotions but also possessing the emotional intelligence to share the responsibility with a colleague, with teamwork. And who has this wisdom in the "heat of the moment?" Not the lone-wolf, that's for sure.

Soon we will discuss further, the importance that collaboration has as a strategy to combat the lone-wolf.

Bullet Proof

It's a funny thing, but I never imagined for a moment that anything would go wrong."

Theo's decision to remove his hands from his insulated gloves was made after weighing up all of the possible scenarios in a heartbeat... in a nanosecond. Any number of choices could have transpired, but they didn't. Theo experienced a lot of shame from that moment. The guilt was etched into his memory. He relived the torment he put Nico through for years, the suffering he caused his family and his employer. Theo would often wake in the middle of the night, sweating, heart racing, reliving the horror, begging to himself, "Why did I do it?"

Eventually, Theo felt he had endured enough. It was time to go into the horror more consciously and discover if there were deeper motives lying beneath his actions, hoping to relieve some of the guilt. During a process of deep contemplation, Theo closed his eyes and recalled the instance when his bare hands were revealed. He freezes that moment in his mind's eye. It was the last time he ever saw his appendages intact, those limbs that served him so well. He visualises their skin colour, their texture, his hair follicles bristling in the sun.

Sinking deeper into meditation and into his feelings, Theo senses the environment, the sounds and smells, his surroundings. He begins to experience his state of mind in that moment. Then slowly a concept of "self" emerges that he hadn't noticed before. "*You know, I completely forgot about something,*" he remembers. "*It's a funny thing, but I never imagined for a moment that anything would ever go wrong.*"

Theo recalls that he felt confident, indestructible. "*You never do something believing you'll get hurt,*" he added. "*In fact, I thought I was a pretty good liney. I needed to beat this nut thing because that's the image I had of myself... bullet-proof.*"

As it turns out, this is actually a pretty common theme. When Theo and I interact with workforces around the world, and they are trusting enough to open up, we often hear things like, "*We are pretty safe already, what's the hullabaloo?*" or "*Workplace safety is too heavy-handed,*" or "*Safety treats us like children.*" Many workplaces reject the notion that anything bad will happen as they begin their daily work tasks.

Actually, we all do in a sense. Believe that nothing bad will happen. What? You've never exceeded a speed limit or made a U-turn at an intersection? You've never negotiated a pedestrian crossing while the animated stick man was still flashing red? Of course you have. Taking risks is normal. It's how I operated when I was on the tools. It's why Theo calls his presentation Just Another Day. You'd be a liar if you said you didn't. Be honest.

Workforces everywhere say they don't deliberately set out to hurt themselves. "They're just accidents," they explain. This point of view is relevant. Because it's basically how things get done. Otherwise, fear and hesitation would reign supreme (like it does sometimes). Theo's revelation softened his guilt, granting him a perspective to view his actions with understanding. This outlook not only brought peace regarding the incident but also unearthed valuable lessons. Ultimately, what's done is done, and the best response is to learn and move forward. Recognising the tragedy, Theo saw an opportunity to guide others from making the same mistake.

Would someone else have acted as Theo did? Past situations hint at its possibility, while Nico didn't it suggests unpredictability. Predicting human responses is impossible. Leaders expecting constant predictability are bound to face setbacks. Our nature compels us to take risks. Suppressing this trait doesn't eradicate it; it merely relocates it to the shadows, lurking, ready to surprise us.

Addressing the role of convenience allows it to transform into a safer alternative. Expecting infallible human behaviour or perfect adherence to the rules is wishful thinking. Until we're substituted by machines, our challenge is navigating and harnessing human tendencies effectively.

Risk & Reward

People place getting the job done, before their own self-preservation and wellbeing, as a natural function of risk and reward.

Being "risky" empowers the human race to progress. Without the "necessary" risks, we wouldn't evolve as a species. We wouldn't leave our front porch to cross the street or mortgage our home on a business idea. Risk and reward are the foundations of capitalism, of all entrepreneurship and enterprise. Pluck and daring are traits we admire and praise, even more than success. Chance and risk are pivotal to the growth and survival of the human race, of any species.

Yet unfortunately, this same entrepreneurial spirit is being crushed in our workplaces every day in the name of safety. Taking risks has become taboo, and it's making our lives a misery. Just look around a pre-shift meeting when the supervisor reads an incident report. That's not concern on the faces of your work crew; it's apathy. Everyone just wants to get back to work. They don't have much tolerance for 'lessons learned' sessions and don't see the value in them. In fact, they'll place getting the job done before their own self-preservation and wellbeing as a function of risk and reward. Risk/reward being a result of pleasure and pain. And pleasure/pain being hard-wired reactions in our bodies that have evolved over the history of our biology, from the dawn of time itself.

Basically, avoiding pain protects organisms against any perceived threats to their integrity or existence. And seeking pleasure entices organisms to sustain themselves. It's fundamental to the theory of Darwinian Natural Selection, the process where organisms better adapted to their environment tend to survive longer.

But it's all perception. It depends on the individual what is risk and what is reward. Picture Theo confidently withdrawing his hands from his gloves. He couldn't see the risk, only the reward (accessing the 12-millimetre nut and finishing the job). The risk/reward equation can appear out of balance because everyone has different points of view, even if it doesn't make safety sense.

Therefore, in any given situation, our decisions in the workplace are drawn towards positive outcomes and repelled by negative ones. But we all assess the benefits and drawbacks, the risks and rewards... differently. This highlights that we never know what a human will do in any given situation. Heightened emotions affect our rationale, and people will react differently depending on their makeup. Remember, the 'Mr. Fix-it' persona that Theo was praised for previously got him into trouble this time. But all this "humanness" is opposite to how we want our safety systems to run. We basically want everyone to (re)act in a certain way, no matter what. Yet I can tell you, it isn't going to happen, whether we apply harsh penalties or not.

We need not "expect" that people will act in a certain way, but to "accept" that we are all irrational human beings. Let's try to avoid safety systems based on expectation and place more emphasis on ramification. If you "know" Convenience causes accidents, you rise above the concept of risk and reward, even pleasure and pain.

Don't imagine that by curbing behaviour you will eliminate accidents either. Or you will forever be at the mercy of behavioural decisions which will leave you dumbfounded, or at least bald from scratching your head in bemused wonderment.

The key to reducing accidents is not to focus on personal behaviour as your benchmark for success, but to meet your workforce right where they are. This means focusing on drivers that motivate people from within their own inspiration. Giving them autonomy and personal responsibility is one way of replacing risk, and recognising individuals who contribute to greater teamwork as the reward.

By providing team members with a deeper awareness of their personal impact on team dynamics, they realise the moment they take their "hands out of their gloves," there is a larger repercussion than just getting the job done. It affects the team. And this is a cause worth noting.

Nature's Way

Those same primal instincts trying to keep you alive—will also get you killed.

Theo attended a safety meeting a couple of days before his accident. His manager asked if anyone knew of any questionable work practices. Theo knew of one. He'd seen a guy working on a live transmission pole who took off their gloves (yep its true). Theo yelled at him to put his gloves back on but received intimidation in response (yep, also true). So in this meeting—Theo said nothing. The guy in question was sitting right behind Theo sweating on what was going to happen. He could feel his energy piercing in his back. It was just an awkward situation. More awkward than raising a serious safety issue unfortunately. This is an example where it is more Convenient to shut up—than speak up. Whereas, safety meetings need to be set up, so that it is more Convenient to say something than be silent. Easier said than done unless you understand the laws of nature.

The pathway of least resistance is physics and Convenience falls within these laws. Knowing the true nature of Convenience is pivotal in avoiding the Convenience Trap. It's worth noting however, that human nature and human behaviour do not operate on the same level. While both are naturally occurring, human behaviour is influenced by the environment whilst seeking Convenience is influenced by our biology. Which means we fall for Convenience when we assume that we can control behaviour. That's the easy assumption. We can't control behaviour, but we can 'inspire' people towards behaving in a way that benefits them.

Humans are very good at following other people's choices if they can rationalise there's something in it for them. We do it all the time. Either consciously or otherwise. If one person steps over a hazard, everyone does. If one person goes around a barrier, they all do. I've seen it over and over again. If we believe someone else has discovered an easier way of doing something, we naturally want to do it too. Herding instincts and mimicking tendencies in humans are natural "tribal" behaviour. And powerful because we are driven by a primitive urge to make things easier to preserve our life. The irony of course is that in the modern workplace, those same primal instincts trying to keep you alive—will also get you killed.

Humans are creatures of habit. Left to our own devices, we will try to turn almost anything into a routine. Because habits allow our minds, thus our bodies, to ramp down more often and conserve energy. Except, habits are relegated as a function of our autonomic system. Subconsciously waiting for a trigger to wake them up. Once a habit is involved, any conscious control is replaced with automatic responses. This is our brain operating at our most basic human level, without much thought at all.

It's understandable we imagine that making things easier is the best possible option. Efficiency, or making the most possible use of the resources at hand, is the single biggest concept on how we run the world. It's what motivated the invention of the wheel. It created the dishwasher, the refrigerator, computers or even something as innocuous as the potato peeler. Inventions to make our lives easier.

But there's a problem if you think these inventions come without a catch. The First Law of Thermodynamics tells us that energy is always conserved. The ratio of energy you save is equal to the energy you spend somewhere else. It is always in unity, a one-to-one ratio. There is always a corresponding consequence for adopting a Convenient choice. You think you are reducing effort but you introduce a new Inconvenience somewhere else. Once you absorb this, you begin to see what's going on around you.

Convenience works everywhere in the world... just not safety. Viewing the workplace from this perspective reveals why we react to certain situations and what is really driving our decisions. You'll realise manifestly that two parts of your "identity" have some explaining to do. They are your *Rational self* and your *Emotional self*: the 'Thinker' and the 'Feeler'. Originally created to preserve us, these *personas* are actually exposing us to danger in the modern workplace!!?

Peculiarly Human

We try to solve a 'feeling' problem with a 'thinking' solution.

Fear and anxiety; pride and hubris; guilt and shame are everyday parts of our lives, located in a part of our brain called the limbic system—'our emotional-self.' It's as if our feelings have some kind of power over us. We can't just turn them on and off whenever we want to. The emotional-self has been an integral part of who we are since the dawn of humankind.

Then, over millennia, another part of our biological identity emerged as an equally dominant force: 'The rational-self'—our neocortex—evolved as a way to temper our emotions and to guide sensible and logical behaviour, our seat of reason. Although animals, specifically mammals, exhibit basic emotional systems, the emotional-self is distinctively human. This part of the brain approaches safety from a perspective that vastly differs from the millions of years of coping with dangers through the emotional-self's freeze, fight or flight responses.

This poses a challenge when implementing modern workplace safety. We find ourselves approaching safety from a mostly rational and logical viewpoint, trying to solve 'feeling' problems (riskiness) with 'thinking' solutions (rules). The problem is, these totally opposite parts of ourselves are continually jockeying for position. The emotional-self tries to run the show while the rational-self keeps things in check. The 'Thinker' creates rules; the Feeler finds ways around them. Both *Selfs* must be considered when designing safety, but they rarely are. The Convenient choice is to create a cognitive-based safety system in spite of the Feeler's emotional leanings. This isn't creating a safe environment at all. We lock the Feeler inside a cage of rules and regulations, hoping the Feeler toes the line. But it won't. Otherwise, our behaviour-based safety programs would have solved all our problems by now.

Historically, Convenience appeared as hunting tools and the domestication of fire, making it easier to eat food. Over many thousands of years, changes to digestion altered our biology and metabolism. But while our bodies continue to tick in Hunter Gatherer time, our exposures in the workplace have taken just decades to evolve. Except, the rational-self is struggling to keep up with these new experiences, which have emerged surprisingly quickly in evolutionary terms.

Today, perceived danger, adrenaline racing through our nervous system at incredible speeds, can be the result of watching a scary movie. Alternatively, cortisol, a component of the body's secondary response to stress, can create anxiety such as speaking in public, rather than confronting a saber-toothed tiger. Dopamine, serotonin, and endorphins, which are gentler and more appealing chemicals, produce feelings of safety, possibly triggered by a financial windfall or watching a feel-good movie. Our biology has repurposed our ancient stress responses for contemporary situations.

When the rational-self can't keep the Feeler in check, it tries to rationalise the apparent 'dysfunctionality' we feel within us—by providing an alter-ego as a reassuring self-image. Hence Mr. Fix-it or the Go-to Guy in Theo's case. Yet, if this self-image cannot be sustained, maybe because it fails to experience the sense of accomplishment the rational-self demands, it is not uncommon for depression or possibly thoughts of suicide to emerge. Thus, suicide seems logical to the rational-self when all 'feels' lost. It throws in the towel as a perfectly rational solution believing that in nature, that which does not serve its purpose makes way for that which does. A totally abhorrent solution for the emotional-self. So, it depends which *Self* you're listening to. It's why humans are prone to repeating errors and maladaptive behaviour. We are designed that way to weed out the flaws in our genes.

Yet evolutionary challenges to our species (humans) are not what they once were. Instincts of self-preservation (fight or flight) are relegated to financial risks, scary movies, or other seemingly benign threats to our existential experience. Anxiety used to be the threat to your survival, now it's going 'all-in' on a poker hand. It's confusing for the *Selfs* to determine what their jobs are anymore... especially in workplaces.

Hard Wired

It's a belief that we take risks and evolve or perish as a species.

It's hardwired in our DNA to transcend fears that would otherwise inhibit evolutionary progress. When people take risks and emerge unscathed, they are rewarded with dopamine. Though I sense it's the thrill (adrenaline) more than the reward sometimes, when the risk seems more appealing than the reward. If you live in a developing country, taking risks is absolutely necessary for your survival, and a sense of pride (dopamine) emerges from being able to put food on the table. So, different cultures are addicted to different chemicals for different reasons.

Both thrill and pride chemicals (adrenaline and dopamine) drive us to take risks, allowing us to achieve things (evolve), which in turn creates more thrill and pride. It's our pleasure centres, always seeking durable fulfilment (convenience), avoiding painful experiences (inconvenience), expending the 'least amount of energy' (survival), that ultimately drive our decisions—an endless natural self-preservation loop.

In any given moment, The rational-self provides a sense of reasoning for the emotional-self's current state. It's a magical partnership derived from biological conservationism. They kind of play off each other. Yet, a couple of key questions emerge. Is it things like pride or stress (and there are others) that drive people to make Convenient choices? Or do people place getting the job done before their own self-preservation for the thrill or adrenaline rush?

The answer is—it's all personal perception. This finely tuned balance of risk and reward is in our heads. It's whatever we believe is the most Convenient option at the time, consuming the least amount of energy, providing the least threat to our existence and, ultimately, creating a personal sense of survival. From this perspective, most people don't realise, but intuitively know, that there is a deeper reason we go to work, beyond just putting bread on the table. 1) We want to feel valued for our contribution, or more importantly, for who we are. 2) We aim to achieve a sense of satisfaction and fulfilment—a fair day's work for a fair day's pay. And 3), we seek to find some meaning in our work tasks, some sense of purpose behind what we do.

Contrary to management's belief, workers place themselves first before the company's interests. As I've mentioned, it's hardwired in our DNA. Humans made it this far because we took risks and sought rewards. No rules or regulations are going to stop that. These natural teleological drivers emerge from an ancient primordial sense to stay the course, propagate our genes, reproduce. It's why a single-cell organism, over millions of years, eventually created multi-celled organisms with consciousness and sensations. It's a remarkable piece of biological engineering suited perfectly for the wilds, but not necessarily the workplace. It's an unconscious belief that we take risks and evolve—or perish as a species. Yet I must reiterate, "unconscious." A deeply ingrained metaphysical construct. We might ask ourselves, "Can we overcome this deeply ingrained belief?" The answer is, absolutely, yes! By bringing it out of unconsciousness. The "Lure of Convenience" can be placed firmly into our awareness through daily reminders, open discussions, and making it part of our safety systems. The rest of this book will now show you how.

Author's Note: OK, let's take a breath. That's the 'biology lesson' over. Thank you for persevering this far. I understand it can be heavy reading. It probably felt more like a science lecture than a book about safety and culture. But it is so important that you appreciate why humans make Convenient choices. That is: "We perceive Convenience will enable us to conserve energy, which ultimately advances us humans as a species." It's so ingrained in us to take shortcuts, and it feels counterintuitive that Convenience could be so detrimental to our wellbeing. The weird thing is that the perils of Convenience are exacerbated when it applies to safety. Seeking Convenience is basically what we do as humans. Once you understand the idea that there are internal natural motivations that get us into trouble in the workplace, we can use these same motivations to combat Convenience-taking in a way that our ego will endorse. If you need to go back and read these chapters again, it will definitely help. But if you totally get it, you're ready to enjoy the rest of this book.

Regards Theo

Control Your 'Selfs'

Once the Feeler is triggered, you either have a strong ally, or a fierce opponent.

In an experiment where participants watched a screen with a cloud of dots moving either to the left or to the right, they were instructed to move a handle in the same direction as the dots. They were good at that until researchers added a directional load to the handle. Making it slightly harder to move the handle in the dots direction. Unconsciously they were influenced by the feeling of least resistance. Regardless of which way the dots were moving.

Stubbornly, the participants swore they moved the handle in the same direction as the dots. But they didn't. Their brains had tricked them. What surprised the researchers was how the participants found it difficult to admit they were wrong. Coming up with all sorts of ways to explain the results. It's a great example of how our Emotional and Rational 'selfs' the

'Thinker' and the *Feeler*—will also come together to maintain a chosen course. If taking a Convenient choice works out, the 'Thinker' is complimentary, *"Well done,"* and the Feeler feels proud. But if a Convenient option fails, the 'Thinker' looks for excuses, *"Oh well, better luck next time,"* and the Feeler is embarrassed or disappointed. These two parts of ourselves cooperate to keep our fragile ego-personas intact.

We see-saw between pride and disappointment, success and failure, good feelings and bad. Opposite ends of our ego going about their merry way. Creating the makeup of our personalities. Sometimes forgiving and patient. Sometimes testy and intolerant. It's the emotional-self trying to expend the least amount of energy while the rational-self justifies our unconscious Convenient-seeking impulses and habits. Except for one small problem. The emotional-self holds the trump card, complete control of your body feelings, which the rational-self hates. If anything, the rational-self is always trying to avoid the emotional-self from hijacking the situation.

Our brains are hard-wired this way. In a constantly revolving unconscious biological process. We don't know that we are even doing it. Until we experience an inner conflict with our Emotional self—the Feeler. Once the Feeler is triggered, you either have a strong ally, or a fierce opponent. Either way, you'll know about it soon enough. Because frustration, anger, worry or doubt will consume you. Either in favour of safety, or against it. Depending on who's side the Feeler is on, Convenience or safety!!? Is it disappointment that the procedures aren't being followed, or because things are taking so long? Right there in that split second, you are going to make a choice that advances your cause or has the potential to derail everything. Including your own wellbeing. In these moments, there's a real possibility that safety could go out the window.

I'm sure you've met people who are so attached to their opinion that they will fight to the death to defend it. Even if they are totally wrong. In a battle for the soul of a person whose emotions are in full swing, it's pointless to try and sway them. You just hope their emotions can maintain some sort of moral conscience and *safety values* kick in. But if the rational-self has done its thing—getting the job done will prevail. The 'Thinker' will do whatever it can to keep the Feeler from throwing a spanner in the works, anything to keep things on track. When the emotional-self refuses to comply, the rational-self cunningly proposes the concept of conserving energy —the ultimate bribe—making the allure of Convenience an irresistibly compelling choice.

Whether these choices are right or wrong doesn't seem to matter. In the *heat of the moment* our only chance of combatting Convenience is to somehow persuade the 'Feeler' to stop *urging* us—a feeling—to hurry up and 'get the job done.' Not an easy thing. In pre-shift meetings or pre-task planning when we are level headed and everything is cool and calm, rational safety makes logical sense. But out in the field we are overcome by an ego-centric world view. We conclude that any type of decision we make, must surely be the right one. Can you identify when you knew something was wrong, but "get the job done" prevailed? If you can't, perhaps the rational-self has a tighter grip on you than you think!

An Inconvenient Truth

We are naive if we expect the controls in the risk assessment to magically appear and slay the Convenience dragon.

When Theo thought about withdrawing his hands from his gloves, he was instantly overcome with morality. *"No Theo, don't. Your training says No!"* But then another conflicting emotion chimes in, *"Wait a minute, I've been doing this for seventeen years and forgotten more than most lineys know."* Buy a car and the same thing happens: 'I love this colour,' followed by, 'Can you afford it?' What ensues is a series of justifications from the rational-self until the deal is sealed—or in Theo's case, his gloves are between his knees.

Therefore, we must provide the emotional-self with safer options to toss around, something that transcends merely appeasing the rational-self's need to accomplish the task. We need to overcome the ego's insistence on being right. Imagine a 'tool' that intervenes at the precise moment you were about to remove your gloves, acting as a saviour. This higher power would keep your ego in check, offering a safer option that doesn't challenge, but motivates you to expend the necessary energy required.

My aim is for you to sidestep the Convenience Trap during work tasks, recognising that under pressure, risk mitigation is often the first victim to go. It's about preempting Convenient choices *before* they occur. The chapter called '*Good Habits Die Hard*' in this book presents practical methods for identifying the Convenience Trap out in the field. Yet recognising there is even a trap at all, will only occur if there is a belief in it.

While I offer various strategies, the effectiveness of any safety ideology hinges on belief. So what I'm about to reveal might push you out of your comfort zone, a common occurrence when you confront the truth. But this isn't about comfort; it's about jolting you out of complacency. Take what you will and leave the rest. The following may not be easy to accept, but I believe it is essential. Let's dive in...

Most workers think safety systems suck. As if you didn't know that already. And if you didn't—you better keep reading. Most people are too afraid to "come out" because criticising safety commitment is sometimes treated like blasphemy. But the controls in the risk assessment aren't going to magically appear and *slay the Convenience dragon* in the heat of the moment. It won't happen like that. Besides, our risk assessments don't regard Convenience as a risk. Maybe they should. Risk assessments and Take-5 pre-task checklists—are in and of themselves—a Convenience. They range from an attempt to transfer the threat of liability, meeting the regulator's whim, to other low-hanging-fruit safety ploys. It's the rational-self attempting to engage the feeling-self's sense of propriety again. And honestly, it doesn't work. If your people had a choice they wouldn't do them. But then, that's also Convenient. So what's the answer?

The 'Inconvenient Truth' is it requires more energy than a piece of paper to be safe. More elbow grease; more discipline; more buy-in... just more! Take-5s are Convenient, and if it's Convenient it's probably bad for you. Anything Convenient is the wrong choice when it comes to safety. It requires much more effort to create a safe-work culture and buy-in of your safety values. So much harder to create belief in the direction and mission of your company. So much more investment to raise morale and put the workers first. We settle for paper-trails to create a sense of personal responsibility, rather than inspiring our workforce with a compelling vision for the future. Or engaging our teams by offering more autonomy and escalating our commitment to openness and disclosure.

Don't misunderstand me; I fully support risk assessments. I absolutely do! I was a Senior Risk Advisor for years. However, risk mitigation done properly. Overcoming the Convenience Trap involves planning work tasks with Convenience in mind and staying vigilant against the mental forces that tempt us towards taking the easy route at every turn. Expecting this level of wisdom to spontaneously emerge in the *heat of the moment*—is bound to disappoint without genuine buy-in from everyone involved. It's crucial not to rely on tools that people don't believe in. It affects morale.

The Performance Trap

Demonstrating how safer decisions can be as convenient as riskier alternatives is key.

Imagine for a moment, in the relentless pursuit of performance, a subtle yet profound reality emerges: the misinterpretation of performance as 'perfection'. This mindset, deeply ingrained in many work cultures, overlooks the hidden reality of workplace dynamics, especially in the face of unforeseen challenges and unpredictable situations—unfortunately, common facts of everyday work life.

Now, picture the concept of Convenience, revealed as the hidden cause of workplace incidents and accidents. By bringing Convenience out of the closet, a comprehension of its implications can lead to transformative insights, enabling workers to reassess their approach to task completion. This reassessment is not just about prioritising tasks differently; it's about redefining what constitutes success in a dynamic work environment.

Our training doesn't just impart skills; it awakens a heightened state of awareness to things otherwise ignored. It's like turning on a light in a room that was always there, yet remained unseen. This new awareness transcends traditional risk assessments, bringing to life a proactive mindset that thrives in the face of the unforeseen. It's about being present, truly present, in every moment of decision.

Performance programs, while designed to enhance productivity and safety, can inadvertently lead to unintended consequences. When safety incidents persist or escalate in the wake of these programs, it indicates a deeper disconnect. It suggests that the focus on performance might be overshadowing the essential human elements of work-flexibility, adaptability, and the freedom to learn from mistakes.

Mistakes are an inevitable aspect of the human experience, particularly in complex work environments. Acknowledging this reality is the first step in creating a culture that can constructively address errors. Our approach to mistakes hinges on a blend of humility and camaraderie, steering away from punitive measures. This strategy is not merely about avoiding discipline; it's about nurturing an atmosphere where learning and growth are integral to everyday work.

Performance, while critical, must be balanced thoughtfully with individual needs like fulfilment, meaning, recognition, and appreciation. Neglecting this balance can lead to a distorted perception where performance is equated with infallibility. Such a scenario can erode the very foundations of safety, productivity, and morale, leading to an environment where the pressure to perform stifles creativity and innovation.

High-pressure scenarios can lead to a perfectionist mindset where teams prioritise task completion, risking a narrow view of performance that sidelines broader role aspects. This approach, often guided by the deceptive allure of Convenience, can create tunnel vision in work crews, mistaking delays for failures. Addressing this requires acknowledging human fallibility and fostering an environment where errors are seen as opportunities for recalibration, not just failures to be penalised. Recognising and addressing these issues is vital when setting performance targets.

Training that fosters humility and collaboration, rather than performance, yields more effective results and higher engagement. It's about creating an ethos where shortcomings are seen as opportunities for learning, not just errors to be corrected. If there is a pattern of neglect or disregard, it may indicate deeper issues—perhaps a lack of faith in your initiatives or a general lack of faith towards safety protocols.

But rarely do performance targets inspire a workforce. They may do so for management, but the workforce is weary of them. 'Emotional' issues cannot be resolved with 'rational' concepts like performance. This is particularly true because it is emotions that drive purpose and mission. The truth is, people don't respond to rational thinking at all; none of us do. We are all irrational, all emotional. Performance may appear to be the option to boost morale, but it never is. Create a great culture, and performance will take care of itself.

Imposter Syndrome

People don't remember there's a piece of paper they've spent hours preparing, sitting in the front of the truck.

Having been involved in countless incident investigations over my career, I've identified the same imposter appearing as a root cause time and time again. Things like: 'Did not identify the risks.' Or sometimes described as: 'Did not follow the defined controls in the risk assessment.' I'm talking about thousands of *incident cause analysis* workshops, all identifying the same root cause. Over and over again. Year after year. It's crazy. What's more, the usual remedial actions are worse. They never change from: 'Must conduct a risk assessment prior to carrying out the task' or 'Must follow the identified controls in the risk assessment.' I'm stuck. So what's really going on here? Is it déjà vu or something more insidious.

When I conduct an incident investigation, I delve deeper

into what the impacted person's state of mind was leading up to the incident. I believe this is more important than following an obviously flawed script. In every case, I found they experienced a brief, almost lightning-fast internal conversation. Debating whether they should take the actual short-cut or not, that ultimately resulted in the incident. They chose the Convenient option because they missed the most important step. They didn't challenge if their choice really was the Convenient route over the "long term." A quick exercise to measure the potential short-term gain against the potential long-term costs may have alerted them towards making a *better* choice. But this wisdom rarely occurs in the field.

We try to make this happen with "Take-5" pre-task checklists or a Job Safety Analysis (JSA)/Task Hazard Analysis (THA). Except, who stops and consults their risk assessment when things heat up? We wish people would, we ask them to, even beg them. Or threaten them with dismissal. But alas, people never remember there's a piece of paper they've spent hours preparing, sitting in the front of the truck. We are human-beings after all I guess, not computers. Maybe we are asking the wrong questions in these tools. We aren't really looking for hazards when things *heat up*, we are seeking the wisdom that avoids Convenience.

What if the Take-5 checklist posed questions like, 'Are you tempted to choose the most convenient option right now?' or 'What will you do if things change in the heat of the moment?' These may seem like rhetorical questions, yet what we truly need to snap us to our senses is a metaphorical slap in the face when we deviate from the plan. A wake-up call, not another rational-based solution like a "Take-5'. If we were to be honest, both workforces and leaders detest Take-5 checklists. They represent our rational thinking-self attempting to make sense of our irrational emotional behaviour. Worse still, Take-5s have evolved into a metric, a KPI. It's no longer about the hazards they identify, but rather, how many we can produce. Do you see now how we repeatedly fall into the rational-self's Convenience Trap time and time again?

Another problem is that humans have a propensity to extrapolate the present and project it into the future. In other words, we tell ourselves that nothing can go wrong, based on current conditions if they stay the same. This creates pride and reinforces our conviction that under current conditions—nothing will go wrong. The more inventive the emotions and rationalisations, the more confident we are in our convictions. However, it's rare for conditions in our workplaces to remain the same. It's kind of irrational when you think about it. Then again, that's the Feeler for you, always looking for ways to be dramatic in the face of rational thinking. Being rational is not the Feeler's job. Therefore, relying on your emotions to be sensible during challenges is not possible.

In reality, the rational-self is a safety imposter. It can't solve our human tendency for taking risks through purely rational concepts. EQ has to be a consideration. This is especially true as it is emotions, not rationale, that affects safe-work decision making. While Convenience might seem like the best option, it's often in resisting the lure of Convenience that we find more innovative, effective, sustainable solutions.

Pathway of Least Resistance

We should be making a concerted effort to reconsider all of the Inconvenient options that we've ignored. The ones where we said, too expensive, too long, too hard.

In nature, efficiency is paramount. The natural world invariably opts for the path of least resistance, whether it's in the realms of physics, biology, or other scientific disciplines. This principle becomes strikingly evident in instances like Theo's unfortunate encounter with a live electrical wire, where he became the easiest conduit for electricity to earth itself. Similar patterns are embedded in our neural pathways. The human brain is conditioned to fulfil necessary tasks by expending the least amount of effort and energy. This extends to sociological interactions, such as seeking validation from colleagues instead of confronting them on dubious safety practices, or avoiding participation in pre-shift meetings to circumvent feelings of unease.

Robert Fritz, an influential figure in personal development, has delved deeply into the structural dynamics that guide human behaviour. Fritz argues that the existing structures in our lives— beliefs, environmental factors, and habitual routines—dictate the trajectory of our actions. By reconfiguring these structures, Fritz contends, one can effectively recalibrate the path of least resistance or in other words, Convenience.

While the notion of the path of least resistance might seem straightforward, it is underpinned by scientific principles that are seldom discussed. Scientists and physicists understand conservation of energy—a foundational concept in physics, aligned with the First Law of Thermodynamics, which postulates that energy can neither be created nor destroyed, only transformed. Every element in the universe that exists, or will ever exist, can be traced back to the initial conditions set forth by the Big Bang. The universe operates on a continuous cycle of transformation, from energy to matter, and then from matter back to energy, ad infinitum.

This brings us to an important consideration regarding workplace safety protocols. Why is it that these often fall short of acknowledging this scientific reality? It appears we are collectively entrapped in a misperception that safety can be enhanced through Convenient, shortcut methods. Risk assessments often factor in elements like budgets, timelines, and available resources. Regrettably, what often transpires is a compromise that disregards Inconvenient yet effective safety measures, viewed as too costly or cumbersome.

A genuine safety protocol should place the onus of responsibility firmly on the individual executing a task. A case in point is the deployment of danger tape to signal hazardous areas. While it may serve as a physical reminder, it is, unfortunately, often mistaken for a control measure, a lamentable misunderstanding considering the tape's inadequacy in averting serious accidents involving heavy machinery.

Current workplace norms such as the mandatory wearing of high-visibility clothing, or the reliance on signage for traffic management, foster a complacency that undermines genuine safety. Such controls engender an illusion of safety where the genuine safeguard should be more awareness from each individual.

What if the prevailing ethos in our workplaces prioritised personal accountability for safety? Supported by an organisational structure that reinforces this paradigm, the impact could be transformative. My associate Theo and I have demonstrated this in practical terms. Workplaces that make personal accountability an overriding principle witness dramatic improvements. When workers own their safety, voluntarily adhering to protocols, the outcome is a workforce empowered by intrinsic motivation.

This innate drive to act responsibly becomes the new path of least resistance, a dynamic recalibration that underlines the efficiency of nature. Once an individual's inner commitment to safety supersedes all else, the default option shifts to an unyielding commitment to safeguarding oneself and others. If you haven't tried it, perhaps now is the opportune moment.

Paying the Price

We all pay a price in our quest for Convenience.
Convenience does not come without a cost.

You'd be amazed at what it takes to make solar panels. The massive consumption of energy and resources might have you reconsider if solar panels are good for the environment. Have you any idea what it takes to make silicone, the core product of solar panels?

Silicone's base component, quartzite, is mined in quarries and trucked to a Smelter; drilling equipment, trucks and machinery consume fossil fuels around the clock; thousands of Hectares of jarrah trees are mowed down annually for the timber which is kilned into massive charcoal logs providing carbon for the smelting process; millions of tons of heavy black smoke pours into the atmosphere; pine trees are ground into wood chips to fuel the chemical reaction between the quartzite and the carbon; 132,000 volts of electricity fuel the submerged-arc furnaces 24 hours a day.

The electrodes are massive. Half a metre in diameter and 3 metres long. The electricity buzzing through them makes a frightening sound. All of these "natural resources" and huge amounts of "energy" burn furiously around the clock. Day after day. Year after year. Meeting the demand for silicone and solar panels. It's good marketing that lulls people into believing solar electricity is "free" and "convenient". It's a false sense of security, if not one of the biggest cons around.

No one talks about these kinds of offsets. They are rarely taken into consideration when we speak in terms of Convenience or conservation. But don't think for a second that I'm anti-solar. Quite the opposite. I'm simply showing you how things can be brought into perspective for a more balanced view. We will reach a "tipping point" where the environmental benefits of solar will far outweigh the environmental costs. I'm certain of it. It's just not happening yet.

Using this analogy, there are many other examples of setbacks in our eternal quest for life's luxuries. Plastic bags, fossil fuel products, fertilisers, chemicals. Not to mention the human costs such as the loss of life at Chernobyl, Bhopal, Deepwater Horizon or the Piper Alpha disaster. There's the impact on the environment, the forests, the sea, the atmosphere and the list goes on and on.

We all pay a price in our quest for Convenience. Convenience does not come without a cost. We always pay a premium. ALWAYS! In safety too.

The core takeaway is the complexity and often overlooked costs associated with Convenience. The production and use of technologies deemed sustainable, such as solar panels highlights a critical aspect of sustainability discussions: the need to consider the entire lifecycle and supply chain of supposedly "green" technologies. Just as we rely on solar panels as a symbol of clean energy, there's a tendency in safety to overly depend on technological solutions or safety devices. This can lead to a false sense of security, where the presence of technology is seen as a complete solution, ignoring the need for ongoing vigilance, maintenance, and human oversight.

This dichotomy raises questions about the true sustainability of such technologies and emphasises the need for a more holistic approach to evaluating "the easier way" impact.

In both scenarios, the long-term impacts can be underestimated. It's the environmental cost of production or specifically with safety, it can be the long-term health impacts of repeated exposure to hazards or the cumulative effect of minor safety breaches.

This analogy extends to other areas such as plastic bags, fossil fuels, and fertilisers, highlighting the broader theme of the hidden costs of Convenience and modern lifestyle choices. Underscoring the importance of a more nuanced view. One that recognises the complexities and trade-offs inherent in our pursuit of Convenience and efficiency.

Real Cost of Convenience

There's a tendency in safety to overly depend on technological solutions or safety devices.

For Theo up there in the work basket, the voices in his head are getting moodier. The rational-self has a problem and is hell-bent on removing that stubborn nut. The emotional-self is backing Theo to solve this tricky situation, and shores up his fighting spirit with doses of adrenaline. Even though it's not lost on Theo, that only last week he was yelling at a guy to put his insulated gloves on. Now he suddenly finds himself considering it. What the...?

But this is going to take some skilful reasoning from the rational-self. Rationalising tough situations is right up its alley. After some persuasion, the emotional-self is eventually on board and applies feelings of frustration to inject some venom. Reinforcing the rational-self's complaints that the job is not getting done. The risk assessment is nowhere to be found. Convenience beckons. A catastrophe looms.

"I wonder," says a voice in Theo's head, *"Perhaps that guy I saw last week is onto something. Maybe gloves-off is the only way of getting at this thing."* To cement the deal, the emotional-self amps things up even more and applies urgency in the form of another impatient body feeling, now fully convinced by the rational-self that this is the best course of action to keep Theo moving in the right direction. Theo considers his training, his experience, what's right and wrong, his pride, his ego, the past, the present and the future—all in a nanosecond.

Most importantly though, he IS NOT weighing up the short-term gains against the potential long-term costs. Of course, the short-term gains are an attractive proposition, up there in the basket. By getting his fingers into that tiny space, there's the Convenience of not having to come up with another plan, Convenience for the crew, Convenience for the customer and for the Company, Convenience to Theo's ego versus a bruised one if he fails. Convenience overrides company values (or at least the expressed ones) and even Theo's own values of personal wellbeing. Convenience overrides everything.

Theo makes a conscious decision to remove his hands from his insulated gloves, perceiving it to be more Convenient than the emotional costs of conceding failure. What of the other possible alternatives then? He could have isolated the power, consulted the engineers or possibly a dozen other solutions? Except it would Inconvenience the crew, the customer, or possibly dent Theo's ego.

Besides, the insulated matting was in place, everything was covered, and Theo knew what he was doing. He had always found the easiest way around things before. And this time he had a model. The guy Theo saw who took his gloves off. Although he'd never seen anyone remove their gloves before, why would Theo even consider it after yelling to this guy to put his gloves back on?

Well, put simply, Convenient options appear too attractive in the *heat of the moment*. Usually weighed against an extremely short time frame. "I am stuck right now, what can I do right now?" But this always exposes us to peril. Particularly because those options have not been weighed against a more accurate time frame. In other words, when you slow things down, what is the potential short-term gain versus the potential long-term cost of this option?

Theo's decision-making process in a high-stakes work environment serves as a powerful metaphor for the broader societal choices we face. It illustrates how short-term Convenience often overrides long-term considerations, potentially leading to catastrophic outcomes. This metaphor resonates beyond environmental issues, touching upon broader themes of human behaviour, risk assessment, and the interplay between rational and emotional decision-making.

What is the real cost of Convenience when the ultimate price is your life?

Wired for Convenience

Taking off our gloves to access tricky nuts,
is our Rational self simply doing its job.

Ironically, we are trained to block out the fear of dangerous work activities, otherwise we wouldn't have the confidence to perform our tasks. Think of skydiving, military exercises, aircraft test pilots, bomb squads, underground miners or even lineys.

For instance, Theo's training involved a unique experience. He was raised in an elevated work platform next to a live transmission line. Theo must reach out and touch the live electrical wire with a metal shifting spanner, wearing insulated gloves. He can hear the tremendous power buzzing through the wire. Every part of his being is saying not to do it.

But this is an important part of a liney's training. If a liney doesn't have the confidence to trust their specialised PPE, they aren't equipped to make calculated decisions as part of their high-risk work activities. Or at least this is what they tell us!

Did desensitising Theo to the risks of high voltage transmission lines contribute to his accident? We'll never know. But habits, either created intentionally or not, have a profound impact on our work tasks.

I was lead investigator of an ICAM (Incident Cause Analysis Method) of a serious incident on an open-pit mining operation in Laos, South East Asia. Nineteen Lao nationals in a convoy of Sadko trucks (Russian personnel carriers) were picking up the remaining drill-crew at the end of the shift. They were parked just twenty metres away from the fully charged drill-hole pattern being prepared for firing. Two men did a final check and started walking towards the trucks when suddenly...

BOOM—BOOM—BANG—BOOM!

The drill and blast supervisor and the shot-firer, safely in the firing shed one kilometre away, initiated detonation thinking that the blast zone was clear. Even after a safety zone clearance sequence was carried out.

Luckily, everyone on the pad was on the upper side of the bench, and the blast percussion blew away from them into the pit. Everyone survived, but they were badly shaken. The root cause of the accident turned out to be *"desensitisation of the blast-zone clearance procedure"*.

Over time, incremental shifts in shot-firing habits had crept in, to save time. What's interesting is that the local national workforce were new to mining. It was the first western-styled mining operation in Laos. They started from scratch. No pre-existing shot-firing sequence prevailed. No existing bad habits. In fact, the locals had a deep respect for explosives. Un-exploded ordinance (UXO) was a massive problem after they were bombed relentlessly during the Vietnam War. Every member of the drill and blast crew could relate a horror story where a family member was affected by UXO. "Live Bomb" remnants litter their villages to this day.

Therefore, the drill and blast crew were trained to a very high standard with a heightened sense of awareness to the dangers of open-pit blasting. Those responsible for the incident could not blame a pre-existing company culture.

The safety sequence in the mining operation was gradually compromised for convenience, a phenomenon I termed "Convenience Kills" even then. My report proposed changes to the drilling and blasting process, including renaming pit locations with local village street names. Bench 3 in the Southern Pit would become "Rue Sam, Huay Xai," to make safety checks resonate more with the Laotian workers.' While the Laotian workers welcomed the renaming of pit areas, Western management was hesitant, missing an opportunity for greater engagement, I felt.

The tendency to seek shortcuts is a natural brain function, especially when tasks become habitual. This instinct leads to safety risks like neglecting blast clearance protocols or removing gloves for tricky nuts. Taking off our gloves to access tricky nuts, is our Rational self simply doing its job. This efficiency of the brain can unwittingly lead to dangerous situations, highlighting our innate inclination towards Convenience.

Right Under Your Nose

We must combat habit patterns relegated to the unconscious mind with alternative trigger and responses.

After repeated exposure to the same risks, we become desensitised to potentially fatal consequences. This is human nature, but still, a huge problem for us. To see what I mean, let's try something. If you focus now, you can see the end of your nose. It's there, take a look. It's always in your view. But our minds have become so efficient at blocking it out we never notice it.

When you decide to pay attention to your nose, your perception changes and you experience another point of view that is closer to reality. This is what is meant by *Convenience Kills*. The idea that there are other options right under our noses which we ignore or overlook in the *heat of the moment*. We naturally avoid looking at our nose like we naturally avoid things that Inconvenience us.

Our brain is always trying to conserve energy and relegate nose awareness as superfluous to our needs. Spacial awareness ignores our nose and places it into our unconscious. Ridding us of the toil and energy it takes to process the information.

This part of our brain has made up our minds for us. Let u s make a wager. I'll bet that you can't maintain focus on your nose for more than a minute while you read this book. Start now, Ok? Such is the power of habit. As far as safety is concerned, we only become alert to the Habitual brain's fight or flight responses when they trigger a body response. Such as a hose that looks like a snake out of the corner of your eye. Or the fright you get from the static you feel when you walk across a wool carpet and touch a metal doorknob. We don't have a conscious choice in any of it. Are you still focusing on your nose or have you forgotten it already? Get it?

Therefore, we must combat habit patterns relegated to the unconscious mind with alternative trigger and responses. That's why a "safety mission" or "culture plan" is so valuable. Or a set of principles that the workforce believe in and buy-in to. These will change existing habit patterns, bringing about awareness to what's really important to your crew, when triggered.

Researchers from the "dots experiment" noted that when we try to decide which apple to pick from a tree, the brain avoids visual stimulus about colour, size and shape, which is the real information required to choose wisely, and instead we focus on the effort required to obtain the apple. We perceive apples at the top of the tree to look less tempting than the lower ones." Dr. Nobuhiro Hagura went on to say,

"Our brain tricks us into believing the low-hanging-fruit really is the ripest."

In the workplace our decisions constantly seek low-hanging-fruit solutions, our perceptions being misled. In fact, it does not make any sense to pick low hanging fruit at all. Typically, if you carry a bag and pick the low hanging fruit first, you would have to work your way up the ladder with an increasingly heavy load.

Therefore, we must brace ourselves in the face of our natural human propensity to avoid "reality" and bring the end of our nose back into focus. You've forgotten about the end of your nose again haven't you?

Given all of the information, how can we possibly know our choices are guiding us in the right direction? How can we stop our emotions, our pride, even our dignity from leading us astray? How can we distinguish our unconscious mind's efforts to make things Convenient, that will ultimately save us energy or kill us? What will steer us in the right direction when it's needed most? Being in the *heat of the moment* feels very 'real' to us. But somehow, we need a process that guides us to make default safety decisions that are reliable, effective and safe. So what to do? What processes can you implement in the workplace today that will have an immediate impact. I'm glad you asked. Would you like to learn it?

The Convenience Culture

We can create a 'team' ethos and triumph over Convenience-related decisions.

The intrigue of the Convenience Trap lies in its double-edged sword; it is both an ally and an adversary. It can streamline processes and enhance productivity, but it also has the potential to erode the foundation of your safety goals. This dichotomy is what makes Convenience a particularly challenging concept for safety professionals to understand. It's not enough to identify Convenience as a causation; there must be a deeper understanding of its implications and the discipline to resist its more perilous temptations. The seduction lies in the very nature of Convenience: It's enticing, it's often equated with efficiency, and it's celebrated in many aspects of modern workplace culture... yet so at your peril!

It is precisely these attributes that make culture shifts away from Convenience so elusive and, consequently, often avoided. Convenience flies under your 'safety radar' because it doesn't present as a direct threat but rather as subtle, incremental actions that erode standards and elevate risk. One effective way to counteract the Convenience culture is by using a real-time safety culture tracking tool, designed to gauge exactly where your work crew stands in terms of falling under Convenience's spell.

As a metaphor, when your people are saying "It sucks" or "I'm stuck," you're likely facing an uphill battle in fostering ownership and responsibility. As you progress, you might hear people acknowledge "I'm good" when they reach the top.

But even better, hearing "We're great" signifies they're making decisions for the team. The Culture Hill powerfully depicts the path that leaders and teams must journey to evolve from being driven by *Convenience,* to one of being guided by a North Star Reference point.

With the right leadership adjustments, we can foster a 'team' ethos, overcoming lone-wolf Convenience-driven decisions and creating a culture where safety becomes collaborative. 'Hilltop Leadership' embodies an approach that avoids blame, excuses, and denial, delivering immediate and tangible results. It empowers a team's safety mindset, encouraging reflection on the language they use and the conduct they demonstrate.

In workplace transformation, leaders and crew members gain an understanding of how their language and conduct significantly impact their personal fulfilment and job satisfaction. Because these attitudes lead to whether people buy-in to your safety message or not. Instead of resorting to blame, excuses and denial, which can lead to dis-ownership, alienation, and conflict, you champion a collaborative approach of 'We Own This—Together!' The results are that lone-wolves begin to form a 'wolf-pack', hunting for Convenient choices.

Mutual support and teamwork can be achieved through training associated with this model. Creating an ownership culture and a workplace where every member experiences the tangible rewards of living above the 'hill'. Out of the shadows of *Convenience* and being led by a *Reference* point, a "why." Our message to anyone about to lead a site meeting of any sort is clear: **Never ask anyone to follow you unless you're at the Top of the Hill.** Unless you want to push 'it' uphill for the rest of the shift.

Leading From the Top of the Hill

From Convenience towards a Reference is Hilltop Leadership.

Your workplace culture isn't what leaders want—it's what teams live. True productivity, quality, and safety hinge on shared commitment, not just compliance. Some teams push for excellence; others settle for less. To lead effectively, you must first see clearly. The Culture Hill model gives leaders real-time insight into how team language and actions shape culture, satisfaction, and results. In the workplace, the words and actions we experience daily are mirroring the culture you've created.

Employee Perception Surveys can provide insight into this collective mindset as does the forty odd years I've spent in lunch rooms and at the lockers. The Culture Hill enhances your awareness of the existing culture, clarifying the link between the type of language spoken, the current cultural disposition, and providing growth opportunities. The signposts on the model depict the OARBED framework: ownership, agreements, respect vs. blame, excuses, denial. They suggest what language and conduct would improve your culture's chances of success. While it also points to what wouldn't. It's not just about avoiding negative behaviours, but actively cultivating a positive, engaged, and respectful working environment.

Consider a leader starting the day with "It Sucks" or "Nothing changes" (I'm stuck) conversations, conspiring with crew members at that level of engagement. That leader will be facing an 'uphill battle' throughout their shift wondering why morale is low or why those who were previously at the top of the hill got sucked into their quagmire of disengagement. It doesn't create job satisfaction or durable fulfilment at all. The lessons from these challenges can be avoided if we realise how potent our conversations are.

You can use the Culture Hill as a tool for leaders to gauge and guide their interactions with their team. It's about understanding how alignment with negativity will hinder your safety standards. Striving for the engagement of "I'm good" and ultimately "We're great" is essential to derive any form of meaning from your workplace. Following is a bit of information around how each zone plays out in the culture:

"It Sucks" represents a culture barely thriving. Typically, these individuals are weeded out. But we all may dip there briefly occasionally. Sustained "It Sucks" behaviour is rare, usually limited to one or two individuals. If pervasive, a separate intervention outside of this scope is advised.

"I'm stuck" signals a significant shift from "It Sucks," but is hampered by indifference and a lack of enthusiasm.

This stage often reflects deep mistrust and can stifle innovation and work pace. It'll also suck you in, so watch out!

"I'm Good" highlights the culmination of the "lone-wolves," focused on individual performance and marked by a desire for personal recognition. It's an improvement from the negativity of previous stages and absolutely required before you can move onto the much more desirable stage. The figure is leaning back and saying "come on" (page 58). It's where your leadership needs to be at a minimum. If your leaders are down the hill you're in trouble. I bet you know what I mean!.

"We're Great" elevates focus from the individual to the team, fostering collaboration and shared success. It's a transformative stage, moving away from the divisiveness of "I'm good" and you're not," the unfortunate side-effect of the lone-wolf, and onwards to a collective achievement and the dedicated eradication of Convenience as an impediment to safety. Using the 'Quick Reference Culture Guide', your instruction is simple: Never ask a team to follow you unless you're "above the hill." Your language and conduct as a leader will ultimately shape your workplace culture.

Row the OAR or Make Your Bed

You've made your BED—now lie in it!

Theo lies in hospital, his arms raised in slings and bound in bandages. Two people enter his ward: the CEO of his company and a Worksafe Inspector, carrying a thick, leather-bound folder. They sit down beside Theo and exchange pleasantries. Theo knows what this is about. The inspector opens the folder, revealing 'Accident Investigation' on the cover. A tense atmosphere descends like a fog.

"Look... stop right there," Theo trembles. "I took my hands out of my gloves. There's nothing to investigate, no one else to blame." With this simple admission, he cuts short the discussion. The case is closed. The inspector quietly closes the file, and the two men take their leave.

Theo wasn't about to play mental gymnastics to avoid the facts—he had done something he would regret for the rest of his life, and now he was facing the music. This unfortunate tale is a sombre reminder of the weight of our choices. After his accident, Theo found himself in the clutches of self-blame, his anguish threatening to consume him. Yet, in the pit of his despair, he discovered a profound truth: "Owning my actions is radically different from blaming myself," he reflects. But how does Theo own something that most people would deem reckless? How does he reclaim agency over such actions and muster the resilience required to reconcile himself with the consequences? By acknowledging his disregard for his own well-being and bravely confronting the memories of his accident, Theo transformed his pain into purpose. He now educates others to avoid the missteps that led to his ordeal. His story illustrates how these principles can be applied in our workplaces, particularly by adopting the Culture Hill combined with the OAR BED framework.

The OAR BED model on page 61 and tied into the Culture Hill on Page 58, has been a cornerstone in my life for over twenty years, which is why it features prominently in the Culture Hill. It's obvious that Ownership, Agreements, and Respect—OAR—will facilitate a rise to the top of the Culture Hill. Just as Blame, Excuses, and Denial—BED—will lead to being stuck at the base. Adopting OAR BED into the workplace is instrumental in fostering personal fulfilment, effective communication, with reinforced commitment, and trustworthiness. It embodies a lifestyle that prioritises self-accountability and integrity.

In the mining, construction, and heavy-industrial theatre, excuses reverberate against the backdrop of your performance. Resigning yourself to 'no choice' is akin to not maintaining your equipment, making you vulnerable to breakdowns. The truth is—we always have a choice. Owning our role in every situation is fundamental to building trust with your team. It is this ownership that can pivot a workplace from playing the blame game to a state where people can significantly influence their personal fulfilment and satisfaction, as well as their workplace's overall enjoyment levels. If we don't enjoy our workspace, odds are it's because we don't trust or believe in it.

Excuses and apologies, deeply rooted in our cultural society, unfortunately serve as a diversion from accountability in the workplace. In heavy industry, where our actions carry significant weight, the habitual use of 'sorry' is debilitating. Neglecting commitments, like habitually arriving late to meetings and dismissing these lapses with apologies, erodes trust—especially if this trait is evident amongst leaders. True ownership acknowledges that these actions stem from personal choice. Acknowledgement that we didn't maintain our agreements is crucial for fostering trust and forming the alliance required of a 'We're great' culture.

This approach shifts the culture from blame and excuses —from 'It sucks' and 'I'm stuck'—to one of proactive engagement—'I'm good' and 'We're great.' Blame and shame are the biggest culture killers of any workplace. Ownership is the saviour. The jargon we use on site captures this well: 'Row your OAR to stay afloat, otherwise or you've made your BED —now lie in it!' *Hmmm*, reminds me of another adage...

When you're pointing your finger, there are three fingers pointing back at you.

Agreements Culture

When you receive an instruction, check in with your higher self to make sure it is something you agree with wholeheartedly. Does it feel right? If not, tweak it; modify it; talk about it; do whatever you need to so that you follow through on that commitment. Or simply go back and renegotiate the agreement until it feels right. Whatever you do, be accountable for carrying out your promises to yourself... as agreed. Trust me you'll be thankful you did. If I had followed my own advice here, I would not have had my accident. It's true!

— Theo Venter

In a world where Convenience overpowers Consequence, Theo's tale highlights that trust and camaraderie are vital for a thriving safety culture. Picture the sterile surgical ward where Theo faces a life-defining decision. The hospital administrator approaches him with a consent form for in case they have to amputate his left arm. Now he knows what that black line is drawn around his bandages. He flat-out refuses, feeling the agreement is being coerced. As anaesthesia takes hold, Theo pleads with the surgeon to save his arm. Upon awakening, he finds Dr. Hanh Nguyen offering a hand, saying, "It's OK Theo, let's shake on it."

Theo's arm survives, embodying the essence of a "real" agreement—a mutual commitment of respect and understanding that persists between Theo and Dr. Nguyen to this day. Agreements foster trust which eliminates the allure of Convenience in subtle ways. How easy to just remove Theo's arm. Agreements at this level are more than a commitment—they're a bond, your word. Work teams who create a culture where agreements are viewed as a pathway to building trust, combat Convenience and shortcuts through team spirit.

Collaboration and mutual respect are the long-term benefits of such work. And as teams move away from Convenience-seeking, they embrace a pathway more fulfilling. Challenges are met with collective resolve and achievements are celebrated as shared victories. In this environment, teamwork flourishes, not just in the adherence to agreements but in the belief in each other's commitment to uphold shared values and goals. Sound lofty? Why don't you try it and see.

Theo's arm, saved by a last-minute agreement, symbolises much more than a physical limb; it represents the strength found in trust, the resilience in facing adversity together, and the power of agreements in forging unbreakable bonds of mutual respect. Through agreements, teams can transcend the ordinary, turning the workplace into a landscape of innovation, accountability, and higher achievement.

How to Make an Agreement?

Begin with a clear, shared objective. Whether it's improving safety measures or enhancing teamwork, understanding what you aim to change is crucial. Focus on collective responsibility. Every agreement should reinforce the notion that together, everyone achieves more. Foster a culture where open dialogue is encouraged, ensuring that every team member's voice is heard and valued. Simplicity is key. An effective agreement is straightforward, measurable, and achievable. Ensure that every agreement reflects the team's core mission and values, reinforcing a sense of shared purpose. Here's the kicker: ***Seal the agreement with a clear understanding of the consequences for not adhering to it****, fostering a culture of accountability. Regular check-ins are vital. Encourage ongoing dialogue to ensure the agreement remains relevant and effective: 'Hey team, any agreements to discuss?'*

How to Keep an Agreement?

When agreements have lapsed, we pinpoint what commitment was missed, focusing on the promise, not the person. A simple 'What was the agreement, did we keep it yes or no' determines if the agreement was upheld, not whether someone is good or bad. We steer clear of shame and focus on the issue: ***Convenience*** *explores the human-ness behind the choice, recognising it as a learning opportunity rather than blame.* ***Consequences*** *are pre-determined if the agreement lapses, emphasising it's a part of the agreement not a punishment. Include alignment with team values, allowing for adjustments to reflect evolving team dynamics. Use setbacks as a stepping stone for new commitments, ensuring they align with team culture goals. Reinforce the consequences of not adhering to the new agreement, emphasising the importance of accountability.*

By embracing Agreements, teams move away from Convenience-seeking (lone-wolf) and embrace a collaborative approach (wolf-pack), where challenges are met with collective resolve. This approach enhances safety performance and builds trust through teamwork, a "We're great" culture.

Good Habits Die Hard

What if we could introduce one simple keystone habit that had the power to transform work crew perception of safety forever? Well... we can!

I don't need to keep harping on about the science of Convenience anymore. You totally get it, yes? The facts are indisputable. You can see hundreds of examples in your own life.

Upon releasing our first book, 'Get Real', Theo and I were astounded by feedback from work crews. A particular chapter, 'Convenience Kills', was having a profound impact. This compelled our attention. Workers expressed pride in resisting shortcuts that might lead to incidents, or worse, accidents or injuries. Thus, we introduced 'The Convenience Trap' into the pre-shift meeting.

These sessions brought to light moments where workers, poised to take shortcuts, recalled 'Convenience Kills' and opted for safer alternatives instead. Consider, for instance, an instance where a worker, rather than hastily changing a water nozzle by crimping the hose, chose to walk back and turn off the tap.

Theo and I were witnessing a paradigm shift. A contest emerged to share the most astonishing Convenience Trap anecdotes. This book springs from these authentic field stories—a testament to real-life safety breakthroughs.

We pondered, what if this singular habit of avoiding Convenience could revolutionise workplace practices across industries?

I reflect on Paul O'Neill's tenure as CEO of Alcoa, starting in 1987. His mission? To make Alcoa America's safest company. Despite financial turmoil and a lack of safety focus, O'Neill's emphasis on safety as a *keystone habit* transformed Alcoa. When he retired thirteen years later, Alcoa's net income had quintupled, and it was the world's safest company.

Consider the dots experiment. If participants had been forewarned about a directional load on the handle, their responses regarding the dots' direction might have been more accurate. Similarly, introducing the Convenience Trap concept in workplaces could have a transformative effect.

I've implemented this at several sites, including BHP's Olympic Dam mine site in Central South Australia. Here, leaders highlighted the dangers of Convenience, making posters of 'Convenience Kills'. Daily, they reminded their teams of this principle, significantly improving safety outcomes. This approach is now a staple in my safety toolkit, consistently yielding positive results.

You, too, should remind your crews daily of Convenience. Observe their reactions when the message resonates. The realisation of this principle often leads to 'Aha!' moments, enhancing awareness of the ever-present Lure of Convenience. As teams report their successes, you'll find that a simple daily reminder, "Convenience Kills," can significantly boost morale. This mantra, echoing in workers' minds before taking shortcuts, has prevented countless incidents and injuries. By understanding and acting on the conflict between what we think is right and what we feel is right, we can navigate towards safer, more effective choices. This is the true essence of the Convenience Trap, a concept with the potential to not only transform workplaces but also to save lives.

Who You Gonna Call?

Should I use this chair or go back to the workshop and get a ladder?"

Long-held beliefs can often be upturned by groundbreaking discoveries. Antonio Damasio, an esteemed neurobiologist, has unveiled a truth that challenges our core understanding: emotions, often relegated to a secondary role, actually play a central role in our high-level cognitive processes. This revelation stands in stark contrast to the twenty-first-century viewpoint that elevates 'reason' above emotion in the intricate workings of the human mind. Interestingly, the famous French philosopher, René Descartes, once declared, "I think, therefore I am." His assertion has influenced our worldview for centuries, advocating reason above all else. Yet, Damasio's research on individuals who had afflictions where they could not generate emotions, has brought us face-to-face with an uncomfortable reality. These people, for all their logical abilities, find themselves paralysed when faced with decisions. This indicates that while we deliberate choices using logic, the final push to "decide" is inherently emotional.

Consider the implications of this in our workplaces. Many organisations, with their strict rules and guidelines, may be overlooking a pivotal component: an emotional connection to safety. This isn't just about dodging hazards; it's about the collective pursuit of satisfaction and well-being that comes from embracing safety as a team. In contrast, Convenience caters to the individual, isolating decision-making from the collective. Imagine a scenario where a worker, under the gun of time constraints, contemplates a choice: opting for a close chair to stand on or making the effort to retrieve a ladder. When safety tools are within easy reach, Convenience aligns with safety. Yet, the crunch of deadlines often glamorises riskier shortcuts as tantalisingly viable, propelled by the urgency to complete tasks by any means necessary. Merely raising awareness falls short; we must evolve into champions of safety, 'Convenience Busters,' who take pride in the safer path over the path of least resistance. The real question is, in the quest for safety, do we lean on the rational self's inclination towards energy preservation, or do we draw from the emotional self's sense of belonging to a team?"

Throughout our lives, we've all felt a common theme: "I feel safe." You never hear "I think safe." Yet, when advising our young or peers, we often emphasise rational thinking. We urge them to "use their heads," though safety's true essence might be emotional: 'follow your heart' or 'trust your gut'. Companies should reshape safety messaging to engage not just the logical, rational mind but also the emotional core of their workforce. This chapter introduces a new paradigm, suggesting we replace traditional safety metrics with "Convenience Busting." Safety isn't just a rational checklist; it's a deeply felt emotion, connecting fundamentally to our well-being and that of others.

The problem with the Thinker's logical-rational approach is that if it's holding a hammer, everything looks like a nail. It treats all safety issues as checklist items. A linear, top-down approach; in series or sequence. In contrast, the Feeler approaches safety like a wedding list, carefully considering who to include or exclude while also addressing the complexity of family dynamics and social status; balancing relationships and expectations. The Feeler embraces human behaviour, emotions, perceptions, and interpersonal complexities, while the Thinker tries to control them. Human behaviour is too irrational for the Thinker.

Convenience Busting is a culture where the safer option is Convenient. Not because it is a rule but because it aligns with a deeper, emotional understanding of safety's true value. This involves re-engineering our environment, policies, and practices to make the right choice instinctive. Rewarding the heart's wisdom over the mind's folly. It's about creating spaces where returning to the yard for a ladder feels more Convenient emotionally, than using a nearby chair, and taking extra safety steps becomes a deeply ingrained habit. In the heat of the moment, *who ya gonna call?* The Thinker or the Feeler?

How To Make the Right Decision

If our values are steeped in 'team' performance, then asking for help is Convenient. If our values honour 'individual' performance, then NOT asking for help is Convenient!!?

In the unforgiving landscape of the savannah, a lioness may coldly cull her litter size against the stark reality of her own chances of survival. Her decisions are devoid of human sentimentality; they're made in the currency of life itself, where the balance of the immediate are weighed against the ultimate cost. She makes her choices based purely on survival, comparing any short-term gains (keeping the cubs) against long-term costs (her own life).

The point I'm making is, that every decision you make is based on what you believe, consciously or unconsciously, is going to give you the greatest advantage over disadvantage in any given moment. It's nature. But in the workplace, humans are constantly making those decisions based on personal ego preferences and values. We aren't that good at make decisions rationally like the lioness. Our safety systems would like to think we can, but our emotional-self has other ideas.

For instance, if our values are steeped in 'team' performance, then asking for help is Convenient. On the flip side, if our values honour 'individual' performance then NOT asking for help is Convenient. Two examples of Convenience with completely different Outcomes. This is worth noting because it teaches us to stop thinking in terms of Convenience, and to start thinking in terms of Consequence.

Take Nico, who concedes defeat in not being able to remove that stubborn nut, passing the spanner to Theo. Theo, gripped by the fear of a bruised ego, refuses to yield, putting his own safety, and that of his team, on the line. Should Theo have elbowed Nico in the ribs up there in the basket and asked for advice? Was that not Convenient? The truth is we may never fully understand the motives behind Theo's choices. So why rely on the whims of Convenience? Imagine if, like the lioness, we stripped away the emotional to reveal the bare bones of decision-making. Unfortunately, we can't. Our decisions are interwoven with our humanness—pride, embarrassment, shame. Our challenge, then, is to find a motive in the face of Convenience, an impulse to rival its lure.

Let's reframe the idea altogether. If our human nature is ego and emotion, which influences our decisions, what can we implement to steer us towards safety? The most effective tool we've discovered is a "Why", a sense of mission, a purpose that resonates with the team's collective values. A workplace aligned with its workers' beliefs ignites a potent intrinsic motivation. Without tapping into these core beliefs, we risk relegating safety to the back burner in pursuit of production and perfection—a place it should never be. The pride in rallying around a shared objective is a formidable motivator, just as shame can be an equally powerful demotivator. These emotions sit at opposite ends of the spectrum, each causing different outcomes in the workplace. While bruised egos can erode morale and precipitate incidents, a spirit of unity fosters buy-in and a drive to do the right thing.

Consider introducing a simple yet pivotal habit that underscores the peril of Convenience. Let the workforce feel the weight of the words "Convenience Kills," and witness the transformation it can spur. The emergence of a robust safety culture can be sparked by such a revelation. There's an unmatched power in an individual who, guided by an "inner voice," chooses safety. This spark of intrinsic motivation is the bedrock upon which we can build a safer environment.

We must then ask ourselves if our safety programs are addressing the right part of the brain. My own experience in drafting countless procedures centred on rationality, compliance, rules and processes, paled in comparison to when I began weaving in the concept of the Convenience Trap. Suddenly, these procedures were met with earnest attention. A workplace that truly understands where its people stand concerning safety will see greater adherence to the principles laid out in this book. We must foster open dialogue, where critique is welcomed, and vulnerabilities are addressed—not in defence of the status quo, but in the spirit of progress and shared commitment to safety.

Investing In Human Capital

Safety has to feel pleasurable, not irritable. Or our pleasure centres will simply resist our safety efforts.

On the surface, one might think that efficiency aligns us more closely with nature, enhancing our resilience. Yet, paradoxically, our pursuit of efficiency has made us more vulnerable. As we've advanced, we've distanced ourselves from the natural world and face new threats: environmental pollution, climate change, mental health crises, and rapid technological growth, to name a few. These modern dangers have outpaced our biological evolution.

The last century has brought technological marvels, but our behavioural evolution hasn't kept the same tempo. The tools and inventions that propel us forward also pose greater risks. As our world becomes smarter and faster, our innate mechanisms for self-preservation lag behind. Despite the passage of time, we seem no nearer to eradicating workplace fatalities than we were a half-century ago.

We cling to archaic terms like "human error" and "violations"—remnants of 1940s safety jargon—because they offer seemingly logical explanations for mishaps. Yet, this is our Rational self, the Thinker talking. Emerging research in human sciences is beginning to challenge these long-held myths, suggesting safety isn't about adherence to rules but is an intrinsic part of our 'makeup'; it's a mindset, a culture, a *feeling* that encompasses our thoughts, actions, and communication.

Herein lies the dilemma: if safe work practices don't resonate with us, if they don't tap into our pleasure centres, we'll default to doing the bare minimum to avoid discomfort. Short-cuts become the norm in the absence of a compelling purpose or the joy that should come from inspiring safety behaviour. Safety, then, is perceived as a tedious obligation rather than a rewarding endeavour.

Safety is due for a renaissance, a fresh perspective that makes it as instinctive and gratifying as any other aspect of our work. It's time for a new approach — a psycho-technology that is able to decipher the Convenience Trap, transforming our engagement with safety from something we do begrudgingly into actions we perform with a sense of pride and satisfaction. Understanding the Convenience Trap isn't just an answer; it's a revolution in how we view and value our well-being at work.

In practical terms, a workplace that integrates the 'Convenience Kills' philosophy is not only likely to see a reduction in accidents and incidents but will also experience a doubling in productivity due to the engagement factor. This boost is not merely a result of heightened awareness but also stems from encouraging people to express any concerns with their levels of fulfilment and satisfaction. Highlighting unsafe practices is just one of the positive outcomes that arise from valuing openness and disclosure. When teams freely share information, regardless of its impact, it not only fosters a culture where poor safety practices are frowned upon but also creates an environment where risks and hazards are actively addressed. The energy injected into work teams from autonomy, drives the effort needed to combat Convenience. If Convenience aims to conserve your energy, then Engagement seeks to utilise your energy fully.

Furthermore, putting 'people first' philosophy will lead to significant cost savings. How? Fewer accidents mean less downtime, reduced insurance premiums, and less money spent on workers' compensation claims. But beyond the financials, the real value lies in the human element—a workforce that feels cared for and respected is so much more productive, engaged, and loyal. And who wouldn't want that?

In closing, adopting the 'Convenience Kills' philosophy is more than a safety strategy; it's an investment in the human capital of a company. It's a commitment to nurturing an environment where safety is as natural and habitual as any other aspect of a job well done. For any workplace looking to not just survive but thrive in today's fast-paced world, this philosophy could be the keystone that ensures longevity, prosperity, and the safety of its most valuable asset — its people.

Long Live Convenience

These principles reside deep within our being, evolving through thousands of years of survival in clans.

I'm sure you are starting to get the picture. Safety has to feel pleasurable, not irritable. Or our pleasure centres will simply resist our safety efforts. Unconsciously or otherwise. Quite often, safety is plain hard work. The main issue is that we have to roll up our sleeves and apply some *elbow grease* to be safe. So the Convenience (pleasure) we seek is countering the effort (pain) it takes. If safety is to win this battle, we must reward safe work practices in a way that celebrates the hard work it takes to be safe. It's that simple.

I know it's ironic, but we need to forget about safety for a moment. Our sense of self-preserve and the conservation of our wellbeing is directly proportionate to the levels of personal satisfaction we have in our workplace. Not the levels of safety. When we *feel* satisfied, safety will take care of itself.

Our work tasks must necessarily adopt a higher ideal than setting boundaries around behaviour. It's not about what we do anyway. It's "why" we do it. It's mission and purpose that get's us to roll up our sleeves and do the right thing. And a sense of pride around doing it.

Coming together to achieve a common goal creates teamwork and a feeling that we are part of a greater cause. This will foster untold engagement and enthusiasm and eliminate the lone-wolf. That's why Mission and Purpose are the single biggest motivators to humankind. These principles reside deep within our being, evolving through thousands of years of survival in clans. It's the Feeler's particular influence again.

Throughout the ages our feelings of self-worth necessarily revolved around pitching in and assisting the tribe. If you didn't pull your weight, you were shunned from the community and your chances of procreation were greatly diminished. A strong incentive to get this right, yes? Therefore counterintuitively, any option you choose that expends the most energy (pain), is ALWAYS the right one. Applying the law of conservation (first law of thermodynamics) means you will always save corresponding energy (pleasure) somewhere else anyway. But you've earned it. You've paid the price. You've got credit in the bank. When you borrow Convenience, you will always have to pay the piper... always!

So don't buy into the illusion that you ever take a shortcut. Or for that matter—a long-cut. You should always be guided by a higher power. A mission, or a sense of purpose. It will provide the impetus to expend the necessary energy and avoid Convenience. A "Convenience Kills" mantra can be the beginning of this process. A *tribal* energy uniting your workforce towards a common cause. If you need a hand to implement this type of culture, the last part of the book will point you our way. But I'm confident you can implement a Convenience-Buster culture just from this book. It's not difficult to grasp.

Therefore, in closing, I'm reminded of the phrase "The King is dead, long live the King." It means the sitting monarch has died and the heir to the throne lives on. As they say, the King never dies. In the same way, Convenience never dies. It is always there, always haunting us. You think you've defeated one murky plan, only to reveal another, and another. Convenience steps right up to meet you, over and over again, time after time. The only defence we have is the awareness, and as crazy as it seems—the "acceptance"—that Convenience is a "real" force to be reckoned with.

When you embrace Convenience, it no longer has power over you and amazingly good options will appear. For instance, I won't stand on this drum, I'll go get a ladder. I won't hit this bolt with my shifting spanner, I'll get a hammer instead. I won't step over this lead, I'll roll it up for others.

Ignoring Convenience or avoiding it altogether is the worse thing you can do. By embracing Convenience and believing that every decision you make will be enticed by its seductive power, means you won't fall victim to potential incidents and accidents. Live and let live. The *Convenience Trap* never dies. Always be aware of its temptation. Or as they say...

"Convenience Kills. Long live Convenience."

When we say "Convenience Kills," it's so workers live.

Moral of the Story

Research shows that Convenience-seeking is steeped in the human condition. Even though there are policies and procedures, protective devices and training to keep us safe, our personas and self-image will override them in our quest for accomplishment.

Our workplaces unwittingly promote Convenience in the name of production and incentives. When companies value individual performance, "lone wolves" seek out Convenience to get the job done. When we value teamwork and camaraderie, "collaborators" seek out advice as being more Convenient. These are two examples with completely opposite outcomes. Yet they are impelled by the same *intrinsic motivation*... the drive for Convenience.

We need to own the part we play as leaders in creating this dynamic. We need to show workers how safer decisions are just as Convenient as risky ones. It's how we train people to catch themselves making Convenient choices in the "heat of the moment" that will avert incidents. So it's up to each of us to take ownership of the roles we play in creating this situation. If we don't own our part in the problem—we can't be part of the solution.

Rapid Cultural Transformation

Convenience Kills Program

Accelerating Safety Success

Your Next Step

Embracing safety as an intrinsic value is the hallmark of an empowered workplace. It's a shift from the old paradigm of "carrot and stick" to one where safety is not an imposed standard but a self-driven commitment. This inner motivation, where satisfaction comes not from external rewards but from the fulfilment of safe practices, elevates the workplace from mere compliance to a culture of proactive discipline.

In such an environment, workers are not merely adhering to safety measures; they are advocates of them, motivated by a personal conviction in the value of safety. This is the essence of the "Convenience Kills" philosophy: it seeks to cultivate a sense of ownership and personal responsibility that transforms safety from a mandate into a way of life.

Under this model, the focus shifts from strict enforcement to inspiring each individual to recognise the subtle yet dangerous allure of convenience—and to choose safety, even when it's not the easiest path. The "Convenience Kills" mantra becomes a beacon, guiding organisations away from the tedium of constant monitoring towards a culture where safety is voluntarily integrated into every action.

Imagine a workplace where safety is so ingrained that reminders become superfluous, where the awareness of convenience's risks is so vivid that every person instinctively upholds safety, owning their actions fully. This vision is not just a theory but a transformative movement.

If you're drawn to this vision, know that Theo and I stand ready to guide you through it. You can find my contact information in the copyright section of the book and listed below. We offer dynamic demonstrations, with Theo's compelling narrative illustrating the dire consequences of ignoring the Convenience Trap. Our workshops are designed not just to inform but to transform, to instigate a profound shift in your organisation's safety culture.

Remember, the most effective safety strategies are those enriched by tangible experiences and actionable knowledge. Our "Convenience Kills" program, combined with the principles of extreme ownership, accountability, and a committed Safety Mission, promises to redefine your approach to safety.

For those ready to champion this change, reach out to us. Detailed in the pages that follow is our full program, ready for your consideration.

Email ken@transfoorm.com or call +61 0415 356 357

Training Program Outline

Convenience Kills™ unequivocally illustrates how Convenient choices are the mysterious root cause of all workplace incidents and accidents. This program begins with a compelling demonstration of how convenience is at the heart of all workplace safety breaches. This revelation, when combined with specialised training, creates an accelerated shift in workforce consciousness and a swift cultural safety transformation. It's currently implemented on the largest mine site in Australia for the biggest mining company in the world, BHP.

Convenience Kills™ takes your workforce on a transformative journey through four distinct cultural stages: a trajectory of emotional and attitudinal change from indifference[1], to questioning[2], then self-direction[3], and ultimately collaborative-spirit[4]. Furthermore, the program counters 'lone-wolf' decision-making, the hallmark of solo efforts, which is the primary cause of Convenient choices leading to accidents. Workplaces struggling to gain buy-in, facing poor operational discipline, or dealing with low morale will gain three pivotal advantages using this hands-on learning experience:

- Comprehensive engagement across the entire department or organisation.
- A significant increase in 'Ownership' amongst leaders and work crew.

- An easy-to-use tracking mechanism that empowers teams to consistently monitor their daily attitudes using a visual cultural-perception tool.

Convenience Kills™ isn't about replacing your current systems; it's about enhancing them. Our program is designed for seamless integration with existing behavioural-based safety initiatives. The unique focus on ownership and personal responsibility fills gaps often overlooked, making the implementation of crucial safety measures not only effective but also more comfortable for those driving the change.

Enduring transformation occurs when individuals commit to a shared philosophy and take personal responsibility for the outcomes. Tangible change arises from personal fulfilment and job satisfaction, rather than the enforcement of behaviour. Convenience Kills™ effectively entwines with your safety goals by overlaying: Clever use of praise and appreciation; Smart ways to value your work crew's contribution; A purpose and alignment with the company's mission and values; The ability to have a voice in decision-making; A feeling and sense of agency and autonomy.

Convenience Kills™ promotes authentic buy-in, transparency, disclosure, initiating support systems and offering shared overarching objectives. When the workforce aligns with a collective belief system, like that of Convenience Kills™, they come together around a recognised standard of behaviour, embracing a mutual safe-work ideology and philosophy. This establishes the foundation for more advanced training, skill development, and the cultivation of innovative ideas. Here are some key aspects:

Unique Perspective on Workplace Safety: The programs central premise that convenience is the root cause of workplace incidents is an intriguing and novel approach. This perspective encourages a deeper understanding of workplace safety breaches, leading to more effective solutions.

Proven Implementation: Its implementation at the largest mine site in Australia for BHP, a major mining company, lends credibility and proves that the program has been tested in challenging environments.

Cultural Transformation: The program promises a transformative journey through four cultural stages, aiming to shift the workforce's consciousness from indifference to collaborative spirit. This gradual and structured approach is effective in fostering lasting behavioural change.

Counteracting Lone-Wolf Decision Making: By focusing on collaborative efforts and countering individualistic decision-making, the program addresses a common cause of accidents, potentially leading to a safer work environment.

Advantages for Struggling Workplaces: For workplaces with issues like poor discipline or low morale, the program offers comprehensive engagement, increased ownership among leaders and workers, and a tracking mechanism for monitoring attitudes. These features are particularly beneficial in revitalising workplace culture.

Seamless Integration with Existing Systems: The program is designed to complement, rather than replace, current safety systems. This aspect will appeal to organisations that already have safety frameworks in place but are looking to enhance them.

Focus on Personal Responsibility and Shared Philosophy: By promoting personal responsibility and a collective belief system, the program aims to foster a more engaged and proactive workforce, which leads to more effective and enduring safety practices.

Promotion of Key Values and Participation: Trust, ownership, openness, and support are emphasised as cornerstone values. Encouraging participation in decision-making and a commitment to continuous improvement will lead to a more dynamic and responsive work environment.

Outcome-Based Approach: The focus on qualitative outcomes, avoiding purely numerical targets, is more effective in creating a safety culture that is intrinsic and value-driven.

In summary, "Convenience Kills™" presents a comprehensive, culturally-focused, and integrative approach to improving workplace safety. Its unique perspective, combined with its structured approach to cultural change and emphasis on personal responsibility and collaboration, makes it an attractive option for organisations seeking to enhance their safety practices and workplace culture.

Theo At Your Workplace

No one knows better than Theo Venter how lethal the Convenience Trap is. Convenience Kills™ is vividly portrayed through his harrowing experience. Your workforce will be transformed as he vividly recounts choosing the Convenience of removing his protective gloves while working on a live power transmission pole, to access a seemingly insignificant 12-millimetre nut. He suffered a massive 22,000-volt shock through his arms and heart. Miraculously, he survived and now shares his story, saving many others from Convenience-related incidents, if not their very lives.

Theo presents Convenience Kills™ to your entire workforce (or specific department), to management, leaders and crew—compellingly showcasing how choices born out of Convenience are the root cause of all workplace incidents and accidents. Theo's real-life ordeal is depicted in this book. Imagine seeing Theo in the flesh on your worksite. Theo's Convenience Kills™ presentation creates a workforce that genuinely grasps and embodies the idea of 'Consequences', especially in relation to ownership.

Research indicates that the human inclination towards Convenience is deeply ingrained. Despite the presence of workplace policies, protective measures, and safety training, this drive often overrides these safeguards, particularly if there are performance and production pressures.

Theo explains how people are motivated by their alter-egos, such as *Mrs Fix-it* or the *Go-to-guy*. These personas bypass precautions in an attempt to complete tasks more efficiently, whether to save time, reduce emotional strain, or cut costs. With the right training however, these personas can also be harnessed to your advantage.

For instance, note that:

> *When you prioritise Individual-Performance, "lone-wolves" will naturally gravitate towards Convenience, to 'get the job done'. On the other hand, when you emphasise Team-Performance, these same individuals see collaboration as the more Convenient option. Both scenarios involve the Lure of Convenience, but yield completely opposite outcomes. Which "performance" model will you foster?*

Introducing the "Convenience Trap" to a workforce offers clear advantages. It fosters a collective understanding and acceptance of a company's or department's safety objectives through a universally recognised principle. Nobody disagrees that whenever they make a Convenient choice during a work-task, it is probably not going to end well. This shared insight forms a mutual understanding, laying the groundwork for follow-up training.

The Allure of "Get the Job Done"

When "Convenience" is understood clearly, you get to witness first-hand the impacts on workplace incidents and accidents. This insight allows workers to consider their vulnerability, particularly when they prioritise task completion above all else, or face unpredictable situations, or unforeseen challenges. Common everyday occurrences. We aim to heighten field awareness during such unexpected scenarios that might not have been discovered in the initial risk assessment. This heightened self-awareness is the primary skill our training instils.

The mindset of "just get the job done" can be seen as a flaw, stemming from a crew's conflation of 'performance' with 'perfection'. When overwhelmed with performance targets, teams might develop a perfectionist mindset. This leads them to focus on task completion as the reward, particularly under pressure. This misinterpretation of performance as merely task completion, has long baffled safety professionals. However, it becomes clearer when you know the source is 'Convenience'.

When implementing performance targets, if safety incidents continue unabated or even increase, unintended drawbacks have arisen from your campaign. Work crews may develop the perception that there's no time to waste, or no room for delays. In the face of unforeseen challenges, what are often mere routine setbacks can be mistaken for failures. It's paramount to illuminate the human aspect of fallibility, granting space for errors and setbacks without the shadow of punishment. Safer options, contrary to belief, don't always tread on borrowed time. The key is to recalibrate perceptions away from believing Convenience 'gets the job done'.

Furthermore, if oversights result from carelessness or disregard, our training provides a method that nurtures humility and avoids punishment. This approach yields superior results and enhances engagement. If sloppiness or neglect exists, it may signal a deeper problem. Possibly doubts in your initiatives or, worse, a general disregard for your safety protocols.

Balancing team-performance with individual performance is crucial. A topic we delve into deeper in the training. If this balance is skewed, the blurred lines between individual-performance, team-performance and perfection can adversely impact safety, efficiency, and worse—team spirit. Simply put, collaboration is safer than solo solutions.

Beyond individual consciousness is the collective mindset, or the culture of your organisation. When the majority of employees undergo this shift in consciousness, your organisation's entire cultural mindset changes. There's a tipping point. Safety becomes a shared value, and the norms and behaviours of your workforce shift away from making Convenient choices hoping to get the job done, instead towards decisions that achieve what needs to be done to make things safe for the team. Not because people "have to", but because they "want to." The journey from Convenience to Performance, that is team-performance as opposed to individual-performance, will be the focus of the training.

Roll-out of the Convenience Kills Program

Leadership and frontline workers must acknowledge the pivotal role they play in promoting Convenience over safety. It's essential to demonstrate to workers that safe choices can be as accessible and appealing as the riskier ones. It's in these decisive "heat of the moment" choices where accidents can be prevented simply by adopting the collaborative spirit.

Theo Venter's poignant "Convenience Kills™" presentation, spanning 1.5 hours including audience interaction at the end, resonates deeply with every member of your team. It's a narrative interwoven with moments of both triumph and tragedy, unequivocally showcasing the dire ramifications of falling prey to the Lure of Convenience, during 'heat of the moment' situations.

Theo's story fosters a profound emotional link to the concept of Convenience. A personal realisation dawns on the audience: that in their work environment, they are consistently drawn to Convenience, a natural human tendency rooted in our biological drive to conserve energy, which often overrides established safety guidelines specifically designed for their protection.

The core objective of the presentation is to establish a foundational belief or a "Keystone habit" that unifies the workforce under one coherent, universally accepted concept: The Lure of Convenience is a normal everyday consideration to be respected like any other risk mitigation technique.

This shared ideology serves as the bedrock upon which further skills and insights are built. Convenience Kills™ provides the following outcomes:

- **Emotional Resonance with Safety**: Theo's presentation fosters a deep-rooted emotional attachment to safety, leading the workforce towards the core tenet of the entire training module – "Ownership". As we progress through subsequent segments of the training, we continually allude to the Convenience Trap, reinforcing its significance in everyday decision-making.
- **Unquestionable Authenticity**: Theo's presentation is impactful to the point where scepticism is rendered obsolete.

Every attendee internalises the message, creating a firm foundation of shared truth. This becomes particularly crucial when addressing companies that face challenges in securing commitment to their existing safety procedures, such as Take-5s and field leadership initiatives.

"Personal ownership of the roles we play in promoting Convenience is essential. Without recognising our contribution to this problem, we cannot be part of the solution." —Theo Venter

Contact admin@transfoorm.com or visit www.conveniencekills.net.

A Case Study - BHP Olympic Dam

An Asset Maximising Its Potential

BHP's Olympic Dam, is the largest mining operation in Australia and reaping significant benefits from the Convenience Kills™ program. As a powerhouse in the mining industry, BHP is always on the quest for innovation and excellence. Convenience Kills™ has been seamlessly integrated into Olympic Dam's surface operations and their flagship performance model, BOS (BHP Operating System).

It's about instilling a new-age safety culture and redefining work team decision-making processes. The tangible results? A considerable reduction in convenience-driven incidents, a reinvigorated workforce, and a robust culture shift towards collective responsibility and ownership. At BHP Olympic Dam, they're not just mining—they're setting industry standards, and Convenience Kills™ is pivotal on this journey.

Here's what they are saying...

When I read 'Convenience Kills' by Ken Roberts and Theo Venter, it echoed my belief that all actions bear corresponding consequences. I shared with Ken that I was walking my children to school and it started to rain. I had an instinct to cross the road shy of the crosswalk to avoid getting wet. Yet, the 'Convenience Kills' core message urges us to use the crossing instead. We see the Convenience Kills™ program imbuing our work teams with this same mindfulness when they conduct their work tasks. The 'green shoots' we are seeing serves as a compelling testament to the effectiveness of this program.

Jennifer Purdie
Asset President, BHP Olympic Dam

The Hilltop Leadership training has significantly influenced me and my team. It unified us, giving direction amid many challenges we were facing culturally. In six months, the progress has been astounding, particularly the follow-up courses which have enhanced my professional and personal skills with improved conflict management and overall work attitude and team spirit.

Anthony Lucisa
Principal Refinery Metallurgist, BHP Olympic Dam

Overseeing numerous departments on Australia's largest mine, it's crucial to have every team operate at its optimum. Ken Roberts and the Convenience Kills™ program have consistently delivered outstanding results across various departments I manage. The transformations I've witnessed are not just temporary "sugar hits" but enduring changes that resonate deeply with our values. I firmly believe in the program's message and Ken's unparalleled expertise. It's why I wholeheartedly recommend that all my managers, and indeed any leader looking for genuine progress, come on board. I firmly endorse Ken's "Ownership" approach. It grounds us in reality.

Sam Francou
GM Surface Operations, BHP Olympic Dam

It's imperative that any initiative we undertake for processing operations aligns with our core principles of BOS (BHP Operating System). The Convenience Kills™ program, especially the Hilltop Leadership model presented by Ken Roberts, seamlessly integrates with our BOS principles. In fact it underpins them. Our superintendents and supervisors have not just adopted the program's mindset, language, and conduct, but they're living it daily, setting an example for everyone. Using Hilltop Leadership in tandem with BOS has revolutionised the way our department operates, creating a synergy that's alive in our day-to-day processes. I can't overstate the positive shift we've experienced. Ken Roberts and Theo Venter are dynamic together.

Kenneth Manyumwa
Manager Processing Ops, BHP Olympic Dam

Having always believed in the potential of my team, it was a revelation to see how Ken Roberts and the Convenience Kills™ program brought out the best in them. With Ken's guidance, we saw an almost immediate uplift in our performance, positioning us as a leading team in the company. It wasn't about changing who we were, but harnessing our collective strength and moving in a unified direction. Ken's expertise and the program's insights were the catalysts, but it was the passion and dedication of my team that truly shone through. They just needed to know they already had it in them, Convenience Kills™ brought it out.

Dave Bojcevski
Manager Technical Processing & Refinery, BHP Olympic Dam

I've always believed that building the right culture is paramount to our organisation's success. Under my tenure I've witnessed our Technical Processing team rapidly embrace transformation and champion our one-team ethos. What sets Convenience Kills apart is its ability to engage our workforce at a deeply personal level. It's not just a safety program; it's a cultural shift that aligns with our values and resonates with our everyday language. This has translated into tangible results, as reflected in our much improved employee perception survey scores.

Cu Phan
GM Surface Development & Planning, BHP Olympic Dam

I can vouch for the profound impact of the Ownership Workshops that I experienced when I was Superintendent TRS Execution . This experience fostered an unparalleled sense of camaraderie and connection within the leader team. The workshop cultivated a shared sense of responsibility and deepening of our commitment to our collective goals. The atmosphere post-workshop was electric and enhanced performance and morale.

Jacob Coles
Superintendent Concentrator Maintenance, BHP Olympic Dam

Convenience Kills™ was a game-changer for our team, and I attribute much of our success to this transformative program. It not only aligned with my safety goals but also empowered me as a leader. It provided practical tools and coaching that helped me guide my team towards a culture of accountability and excellence. Our team's journey saw a significant reduction in incidents and a noticeable improvement in overall safety awareness. Thanks to Ken Roberts and this program, I feel supported in leading my teams.

Ben Adams
Underground Drills Supervisor, BHP Olympic Dam

Hunting the Lone Wolf

Across bustling workplaces, there's a silent saboteur. Picture an eager worker, keen on showing their worth. The currency is task completion. They often become 'lone-wolves', guided by the shimmering light of quick rewards and recognition, they are attracted by the Lure of Convenience to 'get the job done'. "Do I go and get a ladder or should I climb on that drum?" "Do I turn it off at the tap or should I crimp that hose?" Do I take the risk for a moment's triumph?" Every shortcut taken, all in the name of the reward of completion. Any excuse to get it done. . And if it goes wrong, "It was not my fault."

On the flip side, consider a close-knit team, "living above the hill", also keen on showing their worth. Except the currency is camaraderie. Laughter echoes, ideas bounce, and support reigns supreme. Here, the same Convenience takes a different form. It becomes the comforting hand on a mate's shoulder, nudging them to adopt safe practices. It's reaching out from the "top of the hill" to those less believing in the dream. They too complete the task in the same timeframe. Why? Because it's simply easier, and better, for everyone involved to work together towards a goal.

The contrast couldn't be starker. The solo thinker, driven by Convenience, might dazzle briefly but often ends up in the shadows. But there's magic in collective strength. Being a cog in a well-oiled machine, where every member's effort shines and contributes to the team's brilliance, that's the real deal.

Remember, while personal ownership gives you wings, being an invaluable part of 'team ownership' gives you voice. A say in the decisions. Why? Because when the pressure cooker situations arise, your voice is part of the wisdom of the group, and shields everyone from the alluring yet perilous pitfalls of being a 'lone-wolf'.

Conquering the Hill

Imagine standing on a lofty hilltop, the winds of change brushing your face, overlooking the vast expanse of the workplace below. This is where the essence of the Culture Hill thrives, deeply intertwined with the principles of 'Convenience Kills™'. The hilltop is a realm of empowerment, job satisfaction and fulfilment. Above and over the hill, are both a leader and worker's fulfilment.

Alas, picture a leader, starting their day at the bottom of the hill talking "It sucks" and "Nothing changes" (I'm stuck) with the crew. Pushing 'it' uphill like the weight of the world is on their shoulders. Blame, excuse and denial is the currency of their conversation, every step feeling heavier than the last. When a leader is not at the top of the hill before asking the crew to follow them, morning meetings become gruelling hikes, and for the rest of the day it's an uphill slog... culturally. They've set the tone for the level of engagement. Yet, there's no greater teacher than the school of hard knocks. This self-fulfilling prophecy isn't just an abstract concept here. It's a wake-up call to the consequences of a leader's language and conduct. It depends on their attitude, language and ultimately their conduct that sets the crew's tone.

Hilltop awareness is about flipping the 'script' at any given moment in your day. Envision the transformative power of 'Hilltop Leadership'. It's not just about ticking a box for safety protocols. It's about immersing yourself in a philosophy, creating self awareness. And in this awareness, a new language is born—one that resonates with everyone in the organisation. It's about empowering the team and spruiking the benefits of collaboration as the weapon against lone-wolf convenient choices.

It's the chant that unites the workplace, "Convenience Kills"—and let's be honest: every time the seductive Lure of Convenience beckons, there's an inherent risk of dividing the team. Workers intuitively grasp the pitfalls of succumbing to this call if it's always in the back of their mind, especially when the pressure is on, stakes are high, and plans go sideways.

So, we're not just advocating safety—we're championing conscious intent. As challenges loom, our commitment is to keep a vigilant watch, ensuring we act with purpose, especially when the heat of the moment threatens to steer us astray.

The Culture Hill is a practical tool, directly derived from the insights of the Tribal Leadership model by Logan, Fischer-Wright, and King. It meshes seamlessly with OAR BED, offering leaders and crews a clear gauge of their current safety culture. By using this model, you can see the real-time impact of language and behaviour on safety and identify tangible steps for improvement.

This model has its roots in four decades of hands-on experience across mine sites, heavy industrial settings, lunch (crib) rooms and lockers. But remember, it's the teams on the ground that truly define an organisation's culture. Their collective mindset determines everything: the quality of work, the pace of progress, and the benchmarks of success. It's not just about actions; it's the language they use, the stories they tell, and the standards they set that truly shape the fabric of a workplace.

Plotting this conduct on the Hilltop Culture Map allows leaders to discover that their language is reflective of the satisfaction levels of their team. This model was developed directly from monitoring workplaces and the language people use and their conduct. Particularly heavy industrial and mining companies. Whenever there's a confluence of people working side by side, you will discover the same common words appear regularly in hallways and lunch rooms, that reflect the culture. Even Employee Perception Surveys will throw up these same words repeatedly.

Here's to those who see things differently, who challenge the status quo, and who refuse to accept mediocrity. Because those who achieve fulfilment see that lone-wolf is not the answer, but rather the problem. They are the ones who have the power to transform workplaces and cultures. They're the ones who understand that 'I'm good' isn't good enough... it's 'We're great' instead.

We praise the ones who lead from over the top of the hill. Who hunt down Convenient choices in packs. The ones who inspire collaboration, ownership, and a relentless pursuit of 'We're great.' Together, let's all unite and change the world's workplace culture, one step at a time, until the perils of the Convenience-driven mindset is made firmly aware.

Convenience to Reference

Kenneth Manyumwa's leadership at BHP Olympic Dam in South Australia was instrumental in shaping the Culture Hill. Collaborating with his department revealed vital insights that highlighted the need for a structured tool to support the cultural transformation of his Processing Operations. The Culture Hill emerged as a response to this necessity.

Drawing from our studies of *Tribal Leadership**, we discovered that a workplace's cultural journey might begin at any stage, depending on its cultural maturity. We overlaid the idea of graduated cultural stages onto a 'hill' and used terms Kenneth's workforce would actually use. Particularly in

* *Tribal Leadership: Leveraging Natural Groups to Build a Thriving Organisation* by Dave Logan, John King, and Halee Fischer-Wright, which explores the dynamics of workplace culture and how teams evolve through distinct cultural stages.

my interactions, "Nothing's changed" was both a reflection of people's disappointment and also a resistance to change. It emphasised the inertia required to progress from left to right—up and over the hill—from Convenience to Consequence, and from excuses to ownership. This transformative perspective, and the quest for collaboration, turns out to be a driving force behind transforming disengagement into a thriving culture.

The **Sign on the Left of the Hill**: "Blame," "Excuse," and "Denial"

This sign (BED) reflects a culture where individuals avoid accountability and shift responsibility onto others. Typical behaviours include poor attendance, minimal task completion, and resistance to routines. Though this disengagement may appear sporadically in your culture occasionally, it cannot sustain itself indefinitely. This workplace will inevitably self-destruct. Individuals will focus solely on "getting by" without commitment and inevitably stifle growth for themselves and the business. Gauge where your own workplace culture stands by examining the following four stages:

"It Sucks"

Cultures at this stage lack commitment and avoid responsibility, frequently expressing frustration and externalising blame with statements like, "Don't choose me, it's not my area" or "I don't see why we're doing this." Attendance is inconsistent, routines are ignored, and the mindset favours "convenience" over accountability. This stage often reflects a disengaged team where the workplace lacks a solid direction or purpose, leading employees to sidestep responsibility and check out.

"I'm stuck"

At this stage, individuals begin to recognise problems but feel powerless to make meaningful changes, often saying, "Why bother?" or "We've already tried this, it failed then and will fail again." Effort remains minimal, and team members only participate enough to avoid conflict. This stage is characterised by resignation and stagnation, with development progressing slowly or even retreating—as self-motivation remains almost non-existent. Pockets of this type of culture often prevail where performance and safety issues are common, reinforcing a cycle of complacency that makes sustained improvement difficult.

"I'm Good"

At the peak of the hill, this stage often appears to be the primary goal for workplaces on the left-hand side of the cultural journey. It's undoubtedly a significant leap from the resignation seen in the first two stages, yet it has its limitations. This culture centres around personal achievement, with individuals saying things like, 'I'm doing well' or 'Let me hit my targets, and I'll make us look good.' You'll hear statements like, 'Give it to me, I'll do it,' or 'Nobody can do it as well as I can.' This lone-wolf mentality drives excellent attendance and task completion, as people are motivated by their own success. However, it reflects a limited focus on collaboration or shared goals. While the 'I'm Good' mindset is a necessary phase in cultural development and cannot be skipped, it is merely a precursor to the ultimate goal: a unified, collaborative team approach that values inclusivity, openness to feedback, and full engagement.

"We're Great"

At this stage, individuals move from focusing solely on personal success to prioritising team achievement. With phrases like, "We're getting better all the time" or "How can I support the team?", they demonstrate proactive behaviours that uplift others, fostering a collaborative environment where teamwork and shared goals take precedence. In successful workplace cultures, you'll often hear phrases like, "It feels like a family" and "We're all in this together" as part of natural discourse. They've reached this stage through the principles echoed in the sign they are striving for over the hill.

The **Sign on the Right:** "Ownership, Accountability, Respect"

This sign (OAR) signifies a culture where individuals fully embrace collaboration and accountability that says, "We're all for one and one for all" and "Success is everyone's responsibility." Values of ownership, respect, and trust are deeply woven into daily behaviours. People lead by example, support the team, and mentor others, embodying a cohesive and empowered culture. However, sustaining this stage requires an embedded "Agreements Model" that fully integrates the principles of accountability and respect. Only then can the culture evolve into the ultimate vision: a purpose-driven workplace, akin to the Navy SEALs, with the potential for broader impact—perhaps even a global purpose. Without this integration, your culture risks fragmenting back to individual goals, requiring yet another journey up the Culture Hill.

This explanation of the Culture Hill captures what we believe is the journey of transformation from a disengaged culture of convenience to an engaged, accountable team culture that values ownership, collaboration, and respect. With the right training, there is every possibility that your workplace can succeed in traversing the Culture Hill to its eventual **pinnacle of purpose-driven unity**, where individuals are aligned with shared goals and contribute to a culture of collective success and enduring impact. We tried it at one department and then it took off. If you go to Olympic Dam, you will see the Culture Hill depiction everywhere.

Beyond Behaviour

Convenience Kills™ isn't just another safety program—it's a transformative personal development journey tailored specifically to create ownership and accountability. Here's why companies across multiple sectors are tuning in:

Unique Approach

- **Meet People Where They Are**: Our program is designed to cater to individual needs, ensuring maximum resonance and effectiveness.
- **Beyond Behaviour**: We build on the foundation laid by Behavioural Based Safety programs and take a unique approach that focuses on personal language, conduct, and mindset, to ensure your workforce buys into your program.

Transformative Benefits

- **Workplace Dynamics**: Improve perceptions related to inclusion, diversity, and resolve issues of disrespect, bullying, or abuse.
- **Mental and Emotional Wellbeing:** We offer strategic solutions for mental health issues, ensuring a happier, healthier work environment.

- **Culture & Morale**: Transform low morale or poor safety cultures, instilling a sense of purpose and direction.
- **Conflict Resolution**: Address inter-departmental frictions, contractual disputes, and project-based conflicts seamlessly.

Key Program Pillars

- **The Convenience Trap**: Experience the tangible impact of Convenience vs. Consequence, ensuring 100% buy-in.
- **Cultural Ownership**: Transition from a culture of blame to one of ownership, accountability, and response-ability.
- **Hilltop Leadership**: Equip leaders with the skills and mindset to lead effectively, promoting positive safe spaces.

Guaranteed Success

Our confidence stems from:

- **Time-Honoured Principles**: Rooted in values that have stood the test of time.
- **Extensive Experience**: Benefiting from over 45 years in mining and heavy industrial sectors, complemented by 25 years in personal development facilitation.

Your company or department will emerge feeling empowered, with a renewed sense of camaraderie towards their peers and towards management. Discover the Convenience Kills™ difference and elevate your workplace culture to new heights.

Benefits

Deep-Rooted Safety Culture: Rather than just ticking off safety compliance boxes, this approach fosters a culture where every individual deeply cares about and believes in safety. It promotes a proactive environment rather than a reactive one.

Reduced Accidents and Incidents: Recognising and avoiding the Convenience Trap can drastically lower the number of workplace accidents, thus decreasing downtime and associated costs.

Enhanced Employee Morale: When workers believe they are in a safe environment where their well-being is prioritised, their morale and satisfaction levels increase, leading to higher productivity.

Lower Costs: Fewer accidents mean fewer compensation claims, less medical expenses, and decreased equipment repair or replacement costs.

Improved Reputation: Being known as a business that prioritises safety can enhance your reputation among clients, partners, and potential employees, setting you apart in the industry.

Increased Ownership and Accountability: The approach encourages workers to take full ownership of their actions, leading to increased responsibility and decreased reliance on supervision.

Scalable Safety Training: With a core philosophy like "Convenience Kills", safety training can be more consistent, effective, and scalable across different departments or even locations.

Reduction in Enforcement Costs: Less time and resources are spent on policing and enforcing safety guidelines, as workers are intrinsically motivated to follow them.

Long-Term Behavioural Change: This approach seeks to instil a lasting change in behaviour, ensuring safety practices are maintained in the long run and not just adopted temporarily.

Enhanced Decision-Making: By making workers aware of the Convenience Trap, they're better equipped to make decisions that prioritise safety over momentary convenience, even in high-pressure situations.

Higher Retention Rates: A safe and caring work environment can lead to reduced staff turnover, saving costs on recruitment, training, and lost productivity.

Compliance with Regulations: While the approach goes beyond mere compliance, it also ensures that all safety regulations are adhered to, reducing the risk of legal issues or penalties.

Transfoorm

* Email: admin@transfoorm.com or
* Phone: +61 (0)415 356 357

Go the Next Step

If you liked Convenience Kills, then order Get Real today. It is a hard-hitting look at Theo's accident, why a similar accident could happen in your workplace and how to stop it. This book will challenge common workplace customs and practices, providing refreshingly straight talk about what's wrong with the way we think and talk about safety. Uncover the truth about your safety system, even if it flies in the face of your current model of understanding.

People are not your problem... they are the SOLUTION.
Visit: www.theoventer.com/books